Components of System Safety

T0205426

Springer

London
Berlin
Heidelberg
New York
Barcelona
Hong Kong
Milan
Paris
Singapore
Tokyo

Felix Redmill and Tom Anderson (Eds)

Components of System Safety

Proceedings of the Tenth Safety-critical Systems Symposium, Southampton, UK 2002

Safety-Critical
Systems Club

**Data Systems
& Solutions**

Springer

Felix Redmill
Redmill Consultancy, 22 Onslow Gardens, London, N10 3JU

Tom Anderson
Centre for Software Reliability, University of Newcastle,
Newcastle upon Tyne, NE1 7RU

British Library Cataloguing in Publication Data
Safety-critical Systems symposium (10th : 2002 :
Southampton, UK)
  Components of system safety ; proceedings of the Tenth
  Safety-critical Systems Symposium, Southampton, UK, 2002
  1.Industrial safety – Management – Congresses 2.Automatic
  control – Reliability – Congresses 3.Computer software –
  Reliability – Congresses
  I.Title II.Redmill, Felix, 1944- III.Anderson, T. (Thomas)
  1947-
  620.8'6
  ISBN 1852335610

Library of Congress Cataloging-in-Publication Data
A catalog record for this book is available from the Library of Congress.

ISBN 1-85233-561-0 Springer-Verlag London Berlin Heidelberg
a member of BertelsmannSpringer Science+Business Media GmbH
http://www.springer.co.uk

Typesetting: Camera ready by contributors
Printed and bound by the Athenæum Press Ltd., Gateshead, Tyne & Wear
34/3830-543210 Printed on acid-free paper SPIN 10854697

PREFACE

The Safety-critical Systems Symposium has now been held in early February for ten consecutive years, and this book contains the papers presented at the tenth annual event. The Symposium sessions bring together some of the key components of system safety - the investigation of accidents, the definition of safety requirements as well as functional requirements, an understanding of risk, and the recognition of humans and their behaviour as being crucial to safety.

The papers in this book, on these and related topics, are representative of modern safety thinking, the questions that arise from it, and the investigations that result from its application to accident analysis. Nine are written by leading industrialists and five by academics, and all are aimed at the transfer of technology, experience, and lessons, to and within industry. They offer a broad range of views and, not only do they show what has been done and what could be done, but they also lead the reader to speculate on ways in which safety might be improved - for example, through more enlightened management, a systematic application of lessons, process improvement, and a better understanding of risk and how it may be affected by system design. They also indicate new directions in which safety thinking is being extended, for example in respect of information systems whose data is used in safety-related applications, and 'e-safety'.

The papers are presented in the order in which they were given at the symposium and are laid out under six headings that match the symposium sessions:

- Accidents and Their Investigation
- Issues of Low-SIL Systems
- Human Factors
- Safety Requirements
- Risk
- Communication and Electronic Safety

Not only these Proceedings, but those of all ten symposia, have been published by Springer Verlag, and we thank Springer, and in particular Rebecca Mowat, for their supportive partnership during the last decade. But there can be no Proceedings without papers, and we also thank the authors of the papers in this volume for their time and effort, their cooperation, and their responsiveness to our requirements. We also express our gratitude to Data Systems and Solutions for sponsorship of this book of Proceedings.

FR and TA
October 2001

THE SAFETY-CRITICAL SYSTEMS CLUB
sponsor and organiser
of the
Safety-critical Systems Symposium

What is the Club?
The Safety-Critical Systems Club exists to raise awareness and facilitate technology transfer in the field of safety-critical systems. It is a non-profit organisation which cooperates with all interested bodies.

History
The Club was inaugurated in 1991 under the sponsorship of the Department of Trade and Industry (DTI) and the Engineering and Physical Sciences Research Council (EPSRC), and is organised by the Centre for Software Reliability (CSR) at the University of Newcastle upon Tyne. Its Co-ordinator is Felix Redmill of Redmill Consultancy.

Since 1994 the Club has had to be self-sufficient, but it retains the active support of the DTI and EPSRC, as well as that of the Health and Safety Executive, the Institution of Electrical Engineers, and the British Computer Society. All of these bodies are represented on the Club's Steering Group.

What does the Club do?
The Club achieves its goals of technology transfer and awareness raising by focusing on current and emerging practices in safety engineering, software engineering, and standards which relate to safety in processes and products. Its activities include:
- Running the annual Safety-critical Systems Symposium each February (the first was in 1993), with published Proceedings;
- Putting on a number of 1- and 2-day seminars each year;
- Providing tutorials on relevant subjects;
- Publishing a newsletter, *Safety Systems*, three times each year (since 1991), in January, May and September.

How does the Club help?
The Club brings together technical and managerial personnel within all sectors of the safety-critical systems community. It facilitates communication among researchers, the transfer of technology from researchers to users, feedback from users to researchers, and the communication of experience between users. It provides a meeting point for industry and academe, a forum for the presentation of the results of relevant projects, and a means of learning and keeping up-to-

date in the field.

The Club thus helps to achieve more effective research, a more rapid and effective transfer and use of technology, the identification of best practices, the definition of requirements for education and training, and the dissemination of information.

Membership

Members pay a reduced fee (well below a commercial level) for events and receive the newsletter and other mailed information. As it receives no sponsorship, the Club depends on members' subscriptions, which can be paid at the first meeting attended.

To join, please contact Mrs Joan Atkinson at: CSR, Bedson Building, University of Newcastle upon Tyne, NE1 7RU; Telephone: 0191 221 2222; Fax: 0191 222 7995; Email: csr@newcastle.ac.uk

CONTENTS LIST

ACCIDENTS AND THEIR INVESTIGATION

Accident Investigation
– Missed Opportunities

Trevor Kletz

Dept of Chemical Engineering, Loughborough University LE11 3TU

Abstract

After paying the high price of an accident, we often miss the following opportunities to learn from it:

- We find only a single cause, often the final triggering event.

- We find immediate causes but not ways of avoiding the hazards or weaknesses in management.

- We list human error as a cause without saying what sort of error though different actions are needed to prevent those due to ignorance, those due to slips or lapses of attention and those due to non-compliance.

- We list causes we can do little about.

- We change procedures rather than designs.

- We do not help others to learn as much as they could from our experiences.

- We forget the lessons learned and allow the accident to happen again. We need better training, by describing accidents first rather than principles, as accidents grab our attention; we need discussion rather that lecturing, so that more is remembered; we need databases that can present relevant information without the user having to ask for it.

Finally, we ask if legislation can produce improvements.

Introduction

Almost all the industrial accidents that occur need not have occurred. Similar ones have happened before and have been described in published reports. Someone knew how to prevent them even if the people on the job at the time did not. This suggests that here is something seriously wrong with our safety training and the availability of information.

Having paid the price of an accident, minor or serious (or narrowly missed), we should use the opportunity to learn from it. Failures should be seen as educational experiences. The seven major opportunities summarised above are frequently missed, the first five during the preparation of a report and the other two afterwards. Having paid the "tuition fee", we should learn the lessons.

1 Accident Investigations Often Find Only a Single Cause

Often, accident reports identify only a single cause, though many people, from the front-end designers, down to the last link in the chain, the mechanic who broke the wrong joint or the operator who closed the wrong valve, had an opportunity to prevent the accident. The single cause identified is usually this last link in the chain of events that led to the accident.

Just as we are blind to all but one of the octaves in the electromagnetic spectrum so we are blind to many of the opportunities that we have to prevent an accident.

2 Accident Investigations are Often Superficial

Even when we find more than one cause, we often find only the immediate causes. We should look beyond them for ways of avoiding the hazards, such as inherently safer design - could less hazardous raw materials have been used? - and for weaknesses in the management system: could more safety features have been included in the design? Were the operators adequately trained and instructed? If a mechanic opened up the wrong piece of equipment, could there have been a better system for identifying it? Were previous incidents overlooked because the results were, by good fortune, only trivial? The emphasis should shift from blaming the operator to removing opportunities for error or identifying weaknesses in the design and management systems.

When investigators are asked to look for underlying or root causes they may call the causes they have found root causes. One report quoted corrosion as the root cause of equipment failure but it is an immediate cause. To find the root cause we need to ask if corrosion was foreseen during design and if not, why not; if operating conditions were the same as those given to the designer and if not, why not; if regular examination for corrosion had been requested, and if so, if it had been carried out and the results acted upon, and so on. Senior managers should not accept accident reports that deal only with immediate causes.

Most commentators on the disaster at Bhopal in 1984 missed the most important lesson that can be drawn from it: the material that leaked and killed over 2000 people was not a product or raw material but an intermediate. It was not essential to do store it and afterwards many companies did reduce their stocks of hazardous intermediates, often using them as they were made and replacing 50 or more tonnes in a tank by a few kilograms in a pipeline. For ten years since the explosion at Flixborough in 1974, the importance of keeping stocks of hazardous chemicals as low as possible had been advocated. Though reducing stocks saves money as well as increasing safety little had been done. If we can avoid hazards we can often design plants that are cheaper as well as safer.

The report on a serious explosion that killed four men [Kletz 2001b] shows how easily underlying causes can be missed. The explosion occurred in a building where ethylene gas was processed at high pressure. A leak from a badly made joint was ignited by an unknown cause. After the explosion many changes were made to improve the standard of joint-making, such as better training, tools and inspection.

Poor joint-making and frequent leaks had been tolerated for a long time as all sources of ignition had been eliminated and so leaks could not ignite, or so it was believed. Though the plant was part of a large group the individual parts were independent so far as technology was concerned. The other plants in the group had never believed that leaks of flammable gas could not ignite. Experience had taught them that sources of ignition are liable to turn up, even though we do everything we can to remove known sources. Therefore strenuous efforts should be made to prevent leaks and to provide good ventilation so as to disperse any that do occur. Unfortunately the managers of the plant involved in the explosion had little technical contact with the other plants, though their sites adjoined. Handling ethylene at high pressure was, they believed, a specialised technology and little could be learnt from those who handled it at lower pressures. The plant was a monastery, a group of people isolating themselves from the outside world. The explosion blew down the monastery walls.

If the management of the plant where the explosion occurred had been less insular and more willing to compare experiences with other people in the group, or if the directors of the group had allowed the component parts less autonomy, the explosion might never have occurred. The senior managers of the plant and the group probably never realised or discussed the need for a change in policy. The leak was due to a badly made joint and so joints must be made correctly in future. No expense was spared to achieve this aim but the underlying weaknesses in the company organization and plant design were not recognized. However, some years later, during a recession, parts of the group were merged.

The causes listed in accident reports sometimes tell us more about the investigators' beliefs and background than about the accidents.

3 Accident Investigations List Human Error as a Cause

Human error is far too vague a term to be useful. We should ask, "What sort of error?" because different sorts of error require different actions if we are going to prevent the errors happening again [Kletz 2001a].

- Was the error due to poor training or instructions? If so we need improve them and perhaps simplify the task.

- Was it due to a deliberate decision not to follow instructions or recognized good practice? If so, we need to explain the reasons for the instructions as we do not live in a society in which people will simply do what they are told. We should, if possible, simplify the task – if an incorrect method is easier than the correct one it is difficult to persuade everyone to use the correct method - and we should check from time to time to see that instructions are being followed.

• Was the task beyond the ability of the person asked to do it, perhaps beyond anyone's ability? If so, we need to redesign the task.

• Was it a slip or lapse of attention? If so, it no use telling people to be more careful, we should remove opportunities for error by changing the design or method of working.

Blaming human error for an accident diverts attention from what can be done by better design or methods of operation. To quote Jim Reason, "We cannot change the human conditions but we can change the conditions in which humans work."

4 Accident Reports List Causes that are Difficult or Impossible to Remove

For example, a source of ignition is often listed as the cause of a fire or explosion. But, as we have just seen, it is impossible on the industrial scale to eliminate all sources of ignition with 100% certainty. While we try to remove as many as possible it is more important to prevent the formation of flammable mixtures.

Which is the more dangerous action on a plant that handles flammable liquids: to bring in a box of matches or to bring in a bucket? Many people would say that it is more dangerous to bring in the matches, but nobody would knowingly strike them in the presence of a leak and in a well-run plant leaks are small and infrequent. If a bucket is allowed in, however, it may be used for collecting drips or taking samples. A flammable mixture will be present above the surface of the liquid and may be ignited by a stray source of ignition. Of the two "causes" of the subsequent fire, the bucket is the easier to avoid.

I am not, of course, suggesting that we allowed unrestricted use of matches on our plants but I do suggest that we keep out open containers as thoroughly as we keep out matches.

Instead of listing causes we should list the actions needed to prevent a recurrence. This forces to people to ask if and how each so-called cause can be prevented in future.

5 We Change Procedures rather than Designs

When making recommendation to prevent an accident our first choice should be to see if we can remove the hazard – the inherently safer approach. For example, could we use a non-flammable solvent instead of a flammable one? Even if is impossible on the existing plant we should note it for the future.

The second best choice is to control the hazard with protective equipment. preferably passive equipment as it does not have to be switched on. As a last (but frequent) resort we may have to depend on procedures. Thus, as a protection against fire, insulation (passive) is usually better than water spray turned on automatically (active), but that is usually better than water spray turned on by people (procedural). In some companies, however, the default action is to consider a change in procedures first, sometimes because it is cheaper but more often because it has become a custom and practice carried on unthinkingly. Figure 1 (at the end of the paper) describes an example.

6 We Do Not Let Others Learn from our Experience

Many companies restrict the circulation of incident reports as they do not want everyone, even everyone in the company, to know that they have blundered but this will not prevent the incident happening again. We should circulate the essential messages widely, in the company and elsewhere, so that others can learn from them, for several reasons:

- *Moral:* if we have information that might prevent another accident we have a duty to pass it on.

- *Pragmatic:* if we tell other organizations about our accidents they may tell us about theirs.

- *Economic:* we would like our competitors to spend as much as we do on safety.

- *The industry is one: every accident effects its reputation.* To misquote the well-known words of John Donne,

No plant is an Island, entire of itself; every plant is a piece of the Continent, a part of the main. Any plant's loss diminishes us, because we are involved in the Industry: and therefore never send to know for whom the Inquiry sitteth; it sitteth for thee.

7 We Forget the Lessons Learned and Allow the Accident to Happen Again

Even when we prepare a good report and circulate it widely, all too often it is read, filed and forgotten. Organisations have no memory. Only people have memories and after a few years they move on taking their memories with them. Procedures introduced after an accident are allowed to lapse and some years later the accident happens again, even on the plant where it happened before. If by good fortune the results of an accident are not serious, the lessons are forgotten even more quickly. This is the most serious of the missed opportunities and will be considered more fully than the others. [Kletz 1993] describes many examples but here is a more recent one [Anon 2000]:

During cold weather a water line froze and ruptured inside a building. Damage was fortunately not very serious. Three years later the same line froze and ruptured again. The heating in the building was not operating and the water line was near the door. The basement was flooded and two 15 m^3 tanks floated, reached the ceiling and pushed it up by 0.5 m. The incident occurred at a nuclear site. Can we blame the public for doubting the nuclear industry's ability to operate reactors safely when they let the same water line freeze and rupture twice?

The following actions can prevent the same accidents recurring so often:

- Include in every instruction, code and standard a note on the reasons for it and accounts of accidents that would not have occurred if the instruction etc had existed at the time and had been followed. Once we forget the origins of our practices they become "cut flowers"; severed from their roots they wither and die.

- Never remove equipment before we know why it was installed. Never abandon a procedure before we know why it was adopted.

- Describe old accidents as well as recent ones, other companies' accidents as well as our own, in safety bulletins and discuss them at safety meetings.

- Follow up at regular intervals to see that the recommendations made after accidents are being followed, in design as well as operations.

- Remember that the first step down the road to an accident occurs when someone turns a blind eye to a missing blind.

- Include important accidents of the past in the training of undergraduates and company employees.

- Keep a folder of old accident reports in every control room. It should be compulsory reading for new employees and others should look through it from time to time.

- Read more books, which tell us what is old, as well as magazines that tell us what is new.

- We cannot stop downsizing but we can make sure that employees at all levels have adequate knowledge and experience. A business historian has described excessive downsizing as producing the corporate equivalent of Alzheimer's disease [Kransdorf 1996].

- Devise better retrieval systems so that we can find, more easily than at present, details of past accidents, in our own and other companies, and the recommendations made afterwards. We need systems in which the computer will automatically draw our attention to information that is relevant to what we are typing (or reading), as described below.

Of course, everyone forgets the past. An historian of football found that fans would condense the first hundred years of their team's history into two sentences and then describe the last few seasons in painstaking detail. But engineers poor memories have more serious results.

8 Weaknesses in Safety Training

There is something seriously wrong with our safety education when so many accidents repeat themselves so often. The first weakness is that *it is often too theoretical*. It starts with principles, codes and standards. It tells us what we should do and why we should do it and warns us that we may have accidents if we do not follow the advice. If anyone is still reading or listening it may then go on the describe some of the accidents.

We should start by describing accidents and draw the lessons from them, for two reasons. First, accidents grab our attention and make us read on, or sit up and listen. Suppose an article describes a management system for the control of plant and process modifications. We probably glance at it and put it aside to read later, and you know what that means. If it is a talk we may yawn and think, "Another management system designed by the safety department that the people on the plant won't follow once the novelty wears off". In contrast, if someone describes accidents caused by modifications made without sufficient thought we are more likely to read on or listen and consider how we might prevent them in the plants under our control. We remember stories about accidents far better than we remember naked advice. We all remember the stories about Adam and Eve and Noah's Ark far better than all the "dos and don'ts" in the Bible.

The second reason why we should start with accident reports is that the accident is the important bit: it tells us what actually happened. We may not agree with the author's recommendations but we would be foolish to ignore the event. If the accident could happen on our plant we know we should take steps to prevent it, though not always those that the report recommends.

A second weakness with our safety training is that it usually consists of *talking to people rather than discussing with them*. Instead of describing an accident and the recommendations made afterwards, outline the story and let the audience question you to find out the rest of the facts, the facts that they think are important and that they want to know. Then let them say what *they think* ought to be done to prevent it happening again. More will be remembered and the audience will be more committed than if they were merely told what to do.

Jared Diamond writes, "Contrary to popular assumptions cherished by modern literate societies, I suspect that we still learn best in the way we did during most of out evolutionary history – not by reading but through direct experience... For us the lessons that really sink in aren't always those learned from books, despite what historians and poets would like us to believe. Instead, we absorb most deeply the lessons based on our personal experience, as everybody did 5400 years ago." [Diamond 2000]

Once someone has blown up a plant they rarely do so again. at least not in the same way. But when he or she leaves the successor lacks the experience. Discussing accidents is not as effective a learning experience as letting them happen but it is the best simulation available and is a lot better than reading a report or listening to a talk.

9 Reports for Discussion

We should choose for discussion accidents that bring out important messages such as the need to look for underlying causes, the need to control modifications, the need for avoid hazards rather than control them and so on. In addition, we should:

- If possible, discuss accidents that occurred locally. The audience cannot then say, "We wouldn't do anything as stupid as the people on that plant".

- Draw attention to the missed opportunities described above, in particular to the fact that many people have opportunities to prevent accidents. Operators, the last people with an opportunity to prevent an accident, often fall into the traps that others have laid for them but nevertheless often get most of the blame.

- Choose simple accidents. Many engineers are fascinated by complex stories in which someone had to puzzle out some unusual causes. Most accidents have quite simple causes. After a fire, one company gave a lot of publicity to an unusual source of ignition and successfully distracted attention from the poor design and management that allowed four tons of hot hydrocarbon to leak out of the plant. No one asked why it leaked or how the company was going to prevent it leaking again.

Undergraduate training should include discussion of some accidents, chosen because they illustrate important safety principles. Discussion, as already mentioned, is more effective than lecturing but more time-consuming. If universities do not provide this sort of training industry should provide it. In any case, new recruits will need training on the specific hazards of the industry.

10 Databases

Accident databases should, in theory, keep the memory of past incidents alive and prevent repetitions, but they have been used less than expected. A major reason is that we look in a database only when we suspect that there might be a hazard. If we don't suspect there may be a hazard we don't look.

In conventional searching the computer is passive and the user is active. The user has to ask the database if there is any information on, say, accidents involving particular substances, operations or equipment. The user has to suspect that there may be a hazard or he or she will not look. We need a system in which the user is passive and the computer is active. With such a system, if someone is using a word processor, a design program or a Hazop recording program and types "X" (or perhaps even makes a diary entry that there is going to be a meeting on X) the computer will signal that the database contains information on this substance, subject or equipment. A click of the mouse will then display the data. As I type these words the spellcheck and grammar check programs are running in the background drawing my attention to my (frequent) spelling and grammar errors. In a similar way, a safety database could draw attention to any subject on which it has data. Filters could prevent it repeatedly referring to the same hazard.

A program of this type has been developed for medical use. Without the doctor taking any action the program reviews the information on symptoms, treatment, diagnosis etc already entered for other purposes and suggests treatments that the doctor may have overlooked or not be aware of [Anstead 1999].

When we are aware that there is or may be a hazard and carry out conventional searching it is hindered by another weakness: it is hit or miss. We either get a "hit" or we don't. Suppose we are looking in a safety database to see if there are any reports on accidents involving the transport of sulphuric acid. Most search engines will display them or tell us there are none. A "fuzzy" search engine will offer us reports on the transport of other minerals acids or perhaps on the storage of sulphuric acid. This is done by arranging keywords in a sort of family tree. If the there are no reports on the keyword, the system will offer reports on its parents or siblings.

There is ample power in modern computers to do all that I suggest. We just need someone willing to develop the software. It will be more difficult to consolidate various databases into one and to make the program compatible with all the various word processor, design, Hazop and control programs in use.

Chung and co-workers at Loughborough have demonstrated the feasibility of fuzzy searching and carried out some work on active computing [Chung 1998, Iliffe 1998, 1999, 2000].

11 Cultural and Psychological Blocks

Perhaps there are cultural and psychological blocks, which encourage us to forget the lessons of the past.

- We live in a society that values the new more the old, probably the first society to do so. *Old* used to imply enduring value, whether applied to an article, a practice or knowledge. Anything old had to be good to have lasted so long. Now it suggests obsolete or at least obsolescent.

- We find it difficult to change old beliefs and ways of thinking. In the 19th century people found it difficult to accept Darwinism because they has been brought up to believe in the literal truth of the Bible.

- A psychological block is that life is easier to bear if we can forget the errors we have made in the past. Perhaps we are programmed to do so.

The first step towards overcoming these blocks is to realise that they exist and that engineering requires a different approach. We should teach people that "It is the success of engineering which holds back the growth of engineering knowledge. and its failures which provide the seeds for its future development." [Blockley 1980]

12 Can the Law Help?

The Health and Safety Commission have issued *Discussion* and *Consultative Documents* [HSC, 1998, 2001], which propose that companies and other organisations should be required by law to investigate accidents that occur on their premises. Will this reduce accidents? It could but whether or not it will depends on a number of factors:

- Companies could be obliged to follow the follow the first five "opportunities" listed in this paper, but the requirement will be effective only if the Health and Safety Executive (HSE) reads a significant proportion of the reports to see if the underlying causes are found, if suitable recommendations are made and if they are carried out. Will they have the resources to do so? It is easy to tell companies what they should do. It is much more difficult to check that they are doing it thoroughly and not just going through the motions to satisfy the letter of the law.

- Will the law state that its purpose it to investigate accidents *so that action can be taken to prevent them happening again, not to attribute blame?* Industry is coming to realise that many people could have prevented almost every accident but the press and politicians seem more interested in finding someone to blame. (On 18 October 2000, the day after the Hatfield train crash, the *Daily Telegraph*'s front page banner headline was "Who is to blame this time?") When discussing crimes they look for the underlying causes such as poverty, upbringing, ill-treatment and peer-group pressure. In contrast, when discussing industrial accidents, instead of looking for the changes in designs and procedures that could prevent them, they assume they are due to managers putting profit before safety. If the law is interpreted in this spirit it will divert attention away from effective action and will do more harm than good.

  There is a feeling among many people in industry that taken in connection with the proposed legislation on corporate manslaughter the new legislation on investigation of accidents may be used to find culprits to blame. If that proves to be the case the result will be less effective investigations.

  The chief inspector of the Air Accident Investigation Board was quoted in the press as saying that the corporate manslaughter legislation "will adversely affect the culture of our industry, which is for staff and companies to be open about safety concerns" [Smart 2001]. He expresses the fears of many people. I know it is not HSE's intention to encourage a blame culture but its proposals cannot be considered in isolation and the effects of other proposed changes in legislation should be taken into account.

We should remember the words of the report on the tip collapse at Aberfan in 1966 which killed 144 people, most of then children [Anon 1966]: "Not villains, but decent men, led astray by foolishness or by ignorance or by both in combination, are responsible for what happened at Aberfan". The problem is not how to stop bad people hurting other people but how to stop good people hurting other people.

- Will the HSE be able to insist that the reports are published, anonymously if companies wish, so that others can learn from them?

- There is no mention in the HSE *Documents* of the need to see that the lessons of the past are remembered or the methods by which this can be done. Yet unless the information in accident reports is spread and remembered the work involved in the investigation is largely wasted. At best it produces only a local and temporary improvement.

Afterthought

I remember the first time I rode a public bus... I vividly recall the sensation of seeing familiar sights from a new perspective. My seat on the bus was several feet higher than my usual position in the back seat of the family car. I could see over fences, into yards that been hidden before, over the side of the bridge to the river below. My world had expanded. – Ann Baldwin [Baldwin 1995]

We need to look over fences and see the many opportunities we have to learn from accidents.

References

[Anon 1966]   *Report of the Tribunal appointed to inquire into the disaster at Aberfan on October 21st 1966*, HMSO, London, 1996, paragraph 47.

[Anon 2000]   *Operating Experience Summary* No 2000-3, Office of Nuclear and Facility Safety, US Dept. of Energy, Washington, DC, 2000,

[Anstead 1999]   Anstead, M., More needles, less haystack, *Daily Telegraph Appointments Supplement*, 18 November 1999, p. 1.

[Baldwin 1995]   Baldwin, A. D., Letter: *Biblical Archaeology Review*, May/June 1995, p. 50.

[Blockley 1980]   Blockley, D. I. and Henderson, J. R., *Proc Inst Civ Eng*, Part 1, 68:719, 1980.

[Chung 1998]   Chung, P. W. H. and Jefferson, M., A fuzzy approach to accessing accident databases, *Applied Intelligence*, 9: 129, 1998.

[Diamond 2000]   Diamond, J., Threescore and ten, *Natural History*, Dec 2000/Jan 2001, p. 24.

[HSC 1998]   Health and Safety Commission, *A new duty to investigate accidents, Discussion Document*, HSE Books, Sudbury, UK. 1998.

[HSC 2001]   Health and Safety Commission, Proposals for a new duty *to investigate accidents, dangerous occurrences and diseases: Consultative Document*, HSE Books, Sudbury, UK, 2001.

[Iliffe 1998]   Iliffe, R. E., Chung, P. W. H., and Kletz, T. A., Hierarchical Indexing, Some lessons from Indexing Incident Databases, *International Seminar on Accident Databases as a Management Tool*, Antwerp, Belgium, November 1998.

[Iliffe 1999]   Iliffe, R. E., Chung, P. W. H., and Kletz, T. A., More Effective Permit-to-Work Systems, *Proc Safety Env Protection*, 77B: 69, 1999.

[Iliffe 2000]   The Application of Active Databases to the Problems of Human Error in Industry, Iliffe, R. E., Chung, P. W. H., Kletz, T. A., and Preston, M. L., *J Loss Prev Process Industries*, 13:19, 2000.

[Kletz 1993]   Kletz, T. A., *Lessons from Disaster - How Organisations have No Memory and Accidents Recur*, Institution of Chemical Engineers, Rugby UK, 1993.

[Kletz 2001a]   Kletz, T. A., *An Engineer's View of Human Error*, 3rd edition, Institution of Chemical Engineers, Rugby UK, 2001.

[Kletz 2001b]   Kletz, T. A., *Learning from Accidents*, 3rd edition, Butterworth-Heinemann, Oxford, UK, 2001, Chapter 4.

[Kransdorf 1996]   Kransdorf, A., *The Guardian,* 12 October 1996, p. 19.

[Smart 2001]   Smart, K, quoted by Marston, P, *Daily Telegraph*, 27 July 2001.

Acknowledgement

This paper is based on one presented at the Hazards XVI Conference held in Manchester in November 2001 and thanks are due to the Institution of Chemical Engineers for permission to quote from it.

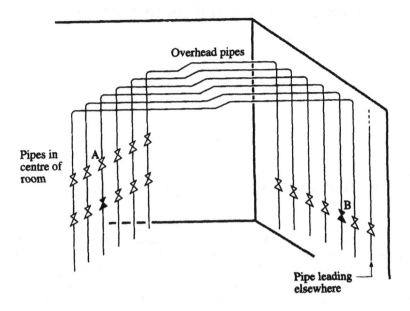

Overhead pipes

Pipes in
centre of
room

A

B

Pipe leading
elsewhere

Figure 1 The fine adjustment valve A had to be changed. The operator closed the valve below it. To complete the isolation, he intended to close the valve on the other side of the room in the pipe leading to valve A. He overlooked the double bends overhead and closed valve B, the one opposite valve A. Both of the valves that were closed were the third from the ends of their rows. Note that the bends in the overhead pipes are in the horizontal plane. When valve A was unbolted the pressure of the gas in the line caused the topwork to fly off and hit the wall.

The report on the incident recommended various changes in procedures. Colour coding of the pipes would have been more effective but was not considered. A common failing is to look for changes to procedures first; to consider changes in design only when changes in procedure are not possible; and to consider ways of removing the hazard rather than controlling it only as a last resort.

The Åsta Train Crash, its Precursors and Consequences, and its Investigation

Gunhild Halvorsrud
Norwegian Railway Inspectorate
Oslo, Norway

1 The Darkest Day

Being as far north as it is, in Norway January is a dark month. In the year 2000, the 4th of January became the darkest day of all. On this day two trains collided head-on on a single-track line at Åsta in Østerdalen, a sparsely populated area north of Oslo. This line, named Rørosbanen, was controlled by a centralised system, but it was not equipped with an automatic train stop (ATS)-system. The line was not electrified, so both trains were diesel-powered. The collision caused the fuel tanks on both trains to rupture and some 5000 litres of diesel poured out and caught fire. The trains burned for several hours. 19 people were killed, some by the collision and some by the fire. Among these were the drivers of both trains. This was the second biggest train accident in Norwegian history. In 1975 a very similar (but for the fire) accident killed 27 people. The train driver survived then, and was prosecuted.

Why did this day become so dark? The direct cause of the accident is still not known, and probably it never will be. It was either an error in the signalling system that thorough investigations have been unable to reveal, or it was simply a signal passed at danger. It doesn't really matter. What matters today is why the accident could happen, and how we are trying to make sure it doesn't happen again. To get an understanding about this, we need to go some 15 years back in time.

2 Background

2.1 Organisational Development

In 1987 the first steps were taken in the process of dividing the national railway company "Norwegian State Railways" (NSB) into a commercially run operating company and a state-owned infrastructure organisation. The reasons for this were twofold:

- The railways had financial problems. By separating the responsibility for the infrastructure from the operation on the lines the railways could be treated just like road traffic.
- EU legislation demanded competition on the track, and a division would make way for that.

NSB was an old institution based on a long tradition of lifelong employment, strong trade unions and a great emphasis on personal responsibility. The manager with responsibility for the safety of traffic was usually recruited from the operational staff, and had a great influence. The engineers worked in big centralised divisions with a strong emphasis on technical safety, whereas the technicians responsible for maintenance were stationed in small local units with a strong feeling of ownership of their section of the line. The work was traditionally rule-based and if something was wrong, it was corrected, with no questions asked about the cost. Gradually this attitude was changed.

First NSB was internally divided into divisions, one of them with responsibility for the infrastructure, one for freight transport and one for passengers. The infrastructure division was then divided into four regions. Each of these was responsible for operation and maintenance in their geographical area. Some of the engineers were then stationed in the regions, some in the main office, and quite a lot of them placed in subdivisions selling their services to the regions in competition with any private enterprises that might want to sell their services. The engineers had to face economic realities. And the engineers got used to them. The success of their subdivisions was dependant on their ability to sell their services. So that became the main goal. The slogan at that time was: More railway for the money.

This was the situation in 1996 when NSB was split into two different organisations. The traffic divisions became a 100% state owned enterprise, but kept the NSB name. The infrastructure manager, Jernbaneverket (JBV) would still report directly to the Ministry of Transport.

The next step was to move most of the technicians in Jernbaneverket into a maintenance division competing with external companies as the engineers had done for the last 6-7 years. They took the changes less easily than the engineers, not being able to see economic success as a satisfying goal. Their motivation decreased and some of them found other jobs.

At the same time The Norwegian Railway Inspectorate was established to safeguard the government's interests in connection with safety issues and to ensure that organisations conducting railway operations comply with the Norwegian railway act and the regulations issued in connection with it. The Inspectorate was intended to have a small staff working at the system level. The main task of the

Inspectorate would be to perform audits, to approve of new and changed infrastructure and rolling stock and to develop legislation in the railway field.

And while all this happened two very important projects had to be carried through.

In 1989 most Norwegians were joyous over the decision to grant the Olympic winter games to Lillehammer, 160 km north of Oslo. It was an excellent location for the games, but the infrastructure capacity was too low. So there was immediate need for action, new roads and improved railways. NSB did splendidly, the infrastructure was ready in time, and during the games, not a single train was greatly delayed. It took lot of work to do this, and even more money, so the backlash was hard. In 1995 and 1996 the budgets were pared to the bone. This affected the maintenance budgets as well as the investment budgets.

At the same time there was a new project. A new airport would be built north of Oslo. The passengers would ride there on the first high speed train in Norway. It would be financed by loans that would be easy to pay back, since the line would become an economic success. (The line is a success, but not economically, but that's a different story). A separate organisation was formed to carry through this work, but the railway competence was brought in from NSB. It was also decided to let the new line have several connections to the old network, in order to make the overall network more effective. The work on the connections was mostly covered by NSB, both regarding manpower and money.

2.2   What Happened to Safety?

The old railway organisation had a great concern for safety. When safety was the issue there was hardly ever any discussion about the cost. Sometimes people took advantage of this and used safety as a lever to get more staff, more money and more pay. When the demand for a more cost-effective railway came, this practise was identified as a problem. The railway was believed to be the safest way of travel, but not the cheapest, not the fastest and certainly not the most reliable. So gradually the focus shifted towards these goals.

Safety was not forgotten, but it was assumed to be there, as a framework. The organisation was still mainly rule based, and it was a common belief that as long as the rules were followed, safety would be maintained. When someone raised their voice and claimed that safety was endangered it was often regarded as another attempt to obtain personal gain.

The workers, who had felt a great personal responsibility for the safety of their own very defined place in the organisation, felt that their place was not so clearly defined any more. They were split from their colleagues into new divisions where the group of people skilled in the different traditional trades were small, and the feeling of being a part of something important decreased. Quite a few reacted to this by finding new employment outside the railways. For the really good ones the pay was better, and the work more interesting. Then they were hired back to do their old jobs, but now their loyalty was elsewhere.

The workers, technicians and engineers didn't really loose interest in safety. But when they saw what little emphasis their leaders put on these issues and how much the concern for financial matters and for punctuality increased, and how important the discussion became on how to save a few seconds here and there, they simply gave up. Of course, the old organisation was not perfect. It was rigid, it was in many ways inefficient, it was not cost effective, but no one had any doubts that safety came first.

2.3   The Work of the Inspectorate

The railway inspectorate was established 01.10.1996. During the first year the number of employees gradually rose to 7. Not until 1999 did the number increase more. The first years were dominated by the task of getting the airport train into action. That involved approval of infrastructure, of rolling stock and of a completely new railway enterprise that were to operate the line. The employees of the Inspectorate are mostly recruited from other industries that have a longer experience with the risk-based safety philosophy than the railways. It took some time and effort during the approval processes to convince all parties that a foundation of risk analysis had to be the basis for an extensive safety management system. But this work succeeded at last. It was worse when it came to influencing the old railway organisations that were not used to having someone supervising them.

This was particularly hard to grasp for Jernbaneverket, which was not put under the supervision of the Inspectorate until February 1998. As a result the first audits ended with critical reports that were not accepted by Jernbaneverket. Some very heated correspondence followed these reports. I don't think anyone really considered the fact that this correspondence was all open to the public, since no one asked for it. Then.

But the first disagreement between Jernbaneverket and The Railway Inspectorate happened earlier, in 1997.

After the split between NSB and Jernbaneverket the responsibility of the rulebook remained in Jernbaneverket. Before 1997 the person responsible for the safety onboard a train had been the chief conductor. The rules stated that both he and the train-driver were to check the exit signals when departing from a station. In 1997 there was a wish to split the responsibility so that the driver would be responsible for the safety of running the train, while the chief conductor would be responsible for the safety of the passengers while entering the train, riding the train and leaving the train. That meant that the conductor would no longer check the signals. This was a sensitive matter, not least amongst the conductors, who felt that they were being deprived of some of their responsibility. It all ended in a discussion whether the conductors were really concerned about safety, or about their own wages.

Jernbaneverket was not under supervision from the inspectorate at that time, but since this was a very important change in the rules, concerning safety, the Ministry of Transport decided that the rules should be approved by the inspectorate. The Inspectorate was sceptical and refused to approve the rules before the effects of the change on safety were determined. This was not done, and the rules were changed without the Inspectorate's approval.

2.4   Was Rørosbanen Forgotten?

After the accident in 1975 when 27 people were killed, probably as a result of a signal passed at danger, plans were made to install an automatic train stop (ATS) system on all lines with centralised traffic control (CTC). ATS is a simpler version of an ATC (Automatic traffic control) system. Its main purpose is to stop a train passing a signal at danger. At that time Rørosbanen was a manually controlled line, with train dispatchers at every crossing station. In the middle of the eighties plans were made to install new signalling systems with CTC on the line.

Building the commonly used interlocking system NSI-63 was not considered cost effective on Rørosbanen. Instead a new concept was developed called NSB-87. The plan was also to install ATS within a few years. But somehow the money allocated to this disappeared, and at the start of the new millennium Rørosbanen was the only line with CTC and without ATS in Norway. At that time the money was reinstated and the ATS installation was re-started.

3   After Day Zero

3.1   The First Report

Seeing all that happened after the accident, it is easy to call the 4th of January 2000 'day zero'. But in the first hectic days after the accident none of us really saw that far. We were shocked by the scenes on our TV-screens showing the wrecked trains and the fire, and hearing again and again the stories of people surviving the crash, but trapped in the coaches, that finally died in the fire.

We heard about people having to leave their friends behind because they had to flee the fire themselves, we heard about the young woman rescued from the train after several hours, having protected herself first with a fire extinguisher and then with a fire hose and we heard the story of the eight young British citizens that saved many of their fellow passengers.

And while we were all trying to digest all this, the internal investigating commission of Jernbaneverket launched their report stating that the cause of the accident was almost certainly a signal passed at danger.

After the accident in 1975 this conclusion was accepted. These kinds of accidents happen, one thought, and the driver was charged. In 2000 this was unacceptable from the public point of view. In 1975 it calmed people down. In 2000 the effect was the opposite. Only five days after the tragedy the last thing the public wanted was the blame to be put on a dead man, unable to defend himself. The last thing they wanted to see was a manager telling them that Rørosbanen was safe. It had just proved not to be. They saw it as an attempt to acquit Jernbaneverket and NSB from blame. The reaction both in the public and the press was close to an uproar. A journalist in one of Norway's main newspaper wrote in a comment to the report (my translation):

"Simply and quickly has the commission made the driver of the northbound train responsible for the great tragedy at Åsta. The commission states that the driver has passed a signal showing red aspect at Rustad station, and therefore the cause of the accident is clear.
This is far fetched. It is a shock to me that Jernbaneverket themselves doesn't see how incomplete the report is"

And closing the article:

"When the internal report of Jernbaneverket makes a deceased train driver responsible for the catastrophe on Rørosbanen, it makes me want to turn against them – and cry with his family and colleagues."

And looking back at all the old articles from those days, I think it was then that the media and the public turned against Jernbaneverket, and all that was said and done by them later only made things worse.

3.2   The First Months

In the months after the accident there were articles in the newspapers almost every day. Looking through them afterwards, I find that I can group them into three main categories:

- Articles about railway safety in general, about the deficiencies in Jernbaneverket and also, but not to the same extent, in NSB.

- Articles about other accidents and near misses on the railways in the last few years.
- Articles about the bereaved and the survivors and their life after the accident.

Already on the 8th of January an independent governmental commission had been established aiming to discover the direct cause as well as the underlying causes of the accident. They conducted a lot of interviews, both concerning the accident and the safety management of Jernbaneverket and NSB. These interviews were mostly held in public with the media, and anyone else who was interested, present. During the course of these interviews, the public gradually developed a picture of the situation both at Rørosbanen and also generally on NSB and Jernbaneverket on the day of the accident. This picture showed some facts:

- Rørosbanen was the only line in Norway with only the driver as a final safety barrier. On all other lines there were either ATS on the exit signal or a train dispatcher present on the platform.
- The two trains were running directly towards each other on the same track for more than 4 minutes before anyone noticed. The control centre was not equipped with acoustic alarm systems that would sound in such situations.
- If such a situation were noticed, the control centre still had no means to stop the trains since they were diesel-powered. Had they been electrical, the power could have been cut off by pressing a button.
- There was no way of making rapid contact with the train crew.

The picture also gave impression of:

- A railway more interested in increasing the capacity on the lines than in safety
- A railway system with opposition between NSB and Jernbaneverket, and also between The Railway Inspectorate and both of them
- A railway which was very rigid and not very open to new ways of thinking
- A railway that did not take the accident as seriously as they should, and kept business as usual.

Of course this picture does not tell the whole truth. It is true that nothing much happened in those months, at least not in the form of concrete actions directed towards the deficiencies of Rørosbanen. Rørosbanen was put back in service on the 13th of February, but not with the CTC active. The line was manned on every crossing station, and manual procedures were put into force.

The independent commission had learned from the experiences of Jernbaneverket, and had hired an independent institution, the Norwegian research and development institute SINTEF, to do a full technical examination of the signalling system at

Rørosbanen. So it was decided to wait for the commissions report before the CTC was put back into service. This also meant that the so dearly wanted ATS system would have to wait. Contracts were not let for delivering acoustic alarms to the control centre, but a group was formed to start work on specifying how it was to be realised.

The acoustic alarm is an example of how different the world looked from the media's point of view than it did from that of the railway experts. It is a common belief that everything can be solved with technology, and even engineers wrote in the press claiming that a buzzer and a relay was all that was needed. Of course it wasn't that simple. It was important to have a system that sounded when there was a real danger, but not at every non-critical error in the signalling system.

Achieving his is quite complicated, especially on single-track lines. And internally in Jernbaneverket there were different opinions about the applicability of these systems. Maybe the money could be better spent on other safety measures? Nobody realised at that time that there was no option.

The first months were quiet. Everyone was waiting, for the independent commission's report and for the police investigation. Then 2 new accidents took place.

4   An "Annus Horribilis"

4.1   An Other Crash

On the night of the 5th of April 2000 a freight train lost its braking power and ran into the back of an other freight train at Lillestøm station some 10 miles east of Oslo. The other train contained two tankers of propane. They were damaged and propane leaked out. After a short time the propane ignited.

The situation was very critical, and investigations following the accident show that we were less than an hour away from a major catastrophe. The fire brigades started pouring water on the tanks to cool them down 2 hours after the crash. A catastrophic explosion would probably have happened ½ to 1 hour later if they hadn't done this. 2000 people living in Lillestrøm were evacuated later in the morning, and it is easy to imagine what could have happened here.

The direct cause of the accident seems to have been an error made by the driver that almost shut the brakes down, but calculations show that the effectiveness of the brakes was overestimated, so the braking distances were set too short. The experts are not sure whether or not correct values would have prevented the accident, but the speed at the moment of impact would have been lower, and the force on the tanks would have been less. But whether they would have ruptured or not is unknown.

4.2   And a Breakdown

New tilting trains were going to be the jewel in the crown for the NSB. They were going to go faster on the existing tracks and would thereby make new and better tracks unnecessary. On the 17th of June 2000 one of these trains derailed at slow speed inside a station area in the south of Norway. The cause of the accident was soon found. One of the axles had broken. Just as in Lillestøm we were lucky this time. If the derailment had happened at full speed on open track, people would have been killed or severely injured. The conclusion of the following investigation was that the axles were too weak for the curved Norwegian tracks. The result was that a speed restriction was placed on the trains until all the axles were replaced with better ones. Later it was established that a permanent speed restriction would be necessary on several of these curves, due to their geometry. And the question arose of how it was possible to overlook such facts when placing an order for 100 million pounds' worth of new trains.

5   The Report

5.1   A Day in November

After all these accidents it is no wonder that the eyes of the press, the public and the railway community were on the independent governmental investigation commission. And we had to wait. The report from the Åsta accident was going to be launched before Easter, then before summer, then in September and it finally came on the 6th of November. The conclusions were as expected, because after all the press had been present at all the interviews. To some there was a great disappointment though, since it reached no final conclusion concerning the direct cause of the accident. It stated:

"The commission cannot with certainty identify the direct cause of the accident that took place on the 4th of January. Neither a signal malfunction nor human error can be excluded. The signal that technically speaking is most likely to have been shown, a red exit signal, is the signal that a driver of a locomotive is least likely to have driven through. Similarly, the least likely signal technically speaking is the green exit signal, the all-clear signal for the driver of a locomotive" [NOU 2000:30]

Still this was perceived a lot differently than the report from Jernbaneverket. The meaning doesn't differ much, but the way it is formulated gives a clearer possibility of a signalling error. And to our great surprise the public took this conclusion as an acquittal of the driver.

The report continues to say that having stated this, it is most important to find the underlying causes of the accident. And the report states:

- ATS was not installed on Rørosbanen
- The departure procedure had been changed without proper authorisation
- Crossing plans had been removed from the driver's timetable on centrally controlled lines. Then the driver wouldn't know whether he was supposed to be crossing an oncoming train at a station or not.
- There was no alarm system in the Traffic control centre. The trains were on a collision course for more than 4 minutes, but no one noticed until less than a minute before the crash.
- There was no communication system for rapid contact between the traffic controller and the driver. When they noticed what was about to happen, the traffic controllers tried to find the mobile telephone number of the trains, but they found the wrong number, and they called the wrong train.

And, perhaps the most important underlying causes:

- No risk-analysis was made prior to changes in priorities, organisation, technical systems and procedures.
- Deficiencies in safety management.

The report concluded by giving a great number of recommendations. They were relating to the causes and included:

- Development of a safety management system.
- Extensive use of risk analysis on all levels of the organisations, especially when planning changes in organisation or technology. The departure procedure should be analysed to see if it is sufficient in all modes of train operation.
- Installation of a safe and reliable communication system on all lines.
- Installation of ATS or the more sophisticated ATC system on all centrally controlled lines.
- Installation of acoustic alarm systems in all control centres.
- Reengineering of the signalling system

5.2 The Signalling System

The result of the SINTEF investigation of the signalling system became an important part of the report. The objective of the investigation was to establish if the signal aspect could have been green for both trains. This work was quite complicated. There was a log on the signalling system, but only the log for the hour before and after the accident was secured. The rest were deleted. With a log for the previous week, which could have been possible to secure at the time, the behaviour of the interlocking prior to the accident would have been easier to determine. The investigation had to be based upon physical investigations, operational tests and on system documentation. No error that could explain the accident was found, but a lot of weaknesses in the system were spotted and a complete reengineering of the system was recommended. And since the system on Rørosbanen was based upon the same principles as the NSI-63 system, the one most widely used in Norway, an investigation into that as well was advised. The commission looked upon these results as so important that they were published prior to the report, in September, so that the work could begin immediately.

This was in many ways the hardest conclusion of all. One of the basic principles for the railways is the fail-safe principle of the signalling system. It just does not fail unsafely. So just as the drivers and their union reacted negatively on the presumption that the driver had passed a signal in danger, the signalling engineers and technicians reacted in disbelief to this recommendation. To many of these experts this was an attack on their own integrity. The concept of reengineering was not very appealing, but as the pressure from the public, fanned by the press, grew, one realised that it could not be avoided. And finally the work was started but actually not until after the report itself was launched.

The reengineering includes both correcting the known deficiencies in the system as well as producing a retrospective safety case built upon the EN 50129 standard. This has proved to be a complicated task.

5.3 After the Report

The report placed the responsibility for the accident on Jernbaneverket as an organisation due to the fact that Rørosbanen was left in a state were one error made by the driver or the signalling system alone could cause an accident like this. The press demanded that general director Steinar Killi should resign, but he was supported by the Minister of Transport who stated that since Killi had only been in office since July 1999, he could not be responsible for what his predecessors had done. But the demand for immediate action was stated clearly by the ministry. The recommendations from the report should be followed up immediately.

The Railway Inspectorate was given the task of supervising the work. The workload placed upon Jernbaneverket was great. The list of recommendations was long and the work was time-consuming and the number of employees in Jernbaneverket that could do the work was limited. The solution was to employ consultants from Norway and the other Nordic countries. The task of seeing that it didn't get out of hand was given to the employees of Jernbaneverket, a task big enough in itself. The schedule they set themselves was very tight; by June 2001 the work should be completed. It turned out to be almost impossible, mostly due to lack of manpower to supervise the great number of consultants.

6   In the Limelight

The railways are not used to being in the centre of public interest. Of course there have been articles in the press on how slow and old-fashioned the railway is, how expensive it is becoming, how bad the service is and so on. But this has always passed. In spite of all the discontent no one had questioned the safety of the railways before. But in an instant it all changed.

In Norwegian legislation there is a demand for great openness in the public administration. This means that unless there is a very good reason preventing it, all correspondence to or from public institutions is open to the public. So when the amount of news directly concerning the accident was diminishing, the journalists started searching the archives to find sources for more articles.

The reason for this was of course that the interest in railway safety issues was still very much alive in the public and news about deficiencies in railway safety sold well. And they found the old reports from the Inspectorates audits, and the correspondence following them, and a picture of a great amount of disagreement between the Inspectorate and Jernbaneverket emerged. The picture was worse than reality, mostly because of the colourful language used in some of the letters. And this really sold!

Why did these news become so exciting? In my opinion there are three reasons. The first is the obvious one, that disagreements are exciting. The second is the opinion in the public that Jernbaneverket was trying to free itself from blame by blaming the driver. He was looked upon as the underdog, the scapegoat for a large and powerful organization. And he was dead; he could not defend himself. The public found this very unfair and welcomed all attempts to place the blame elsewhere. The third reason is the former belief that the railways, despite all their other deficiencies, were safe.

The other accidents that Spring only contributed to keeping the pressure up. By the end of the summer I was really wondering how anyone not connected directly with the railway could possibly want to read another article about all this mess. And for everyone connected to the railways in any way the strain of all this attention, mostly negative, was becoming very large. People reacted differently. Some just didn't care, some developed quite a rage against journalists and others became ill.

When the independent commission's report came out it all started again. But now the organisations were better prepared. There were press releases welcoming the report, stating how valuable it would be in the work of improving the organisation. Action plans were made, tasks were assigned. There was no time to sit down and discuss, to analyse the relative importance of the recommendation, what to do first and last. There was no time to consider if there were any other actions that could have had the same or better effect with less effort. And if there had been, there had to be very good reasons for it, and they had to be explained in a very simple way, or the public would not have accepted them. By that time the public's trust in Jernbaneverket was so low that no one would have listened anyway.

The cost then, of course, became a lot higher than it could have been. To be able to carry all the work through there had to be consultants hired, lots of them, at any cost. Tasks that would have been better performed in series, had to be carried out in parallel. And because of the hurry, errors were made and some of the work had to be done over again.

How could the organisation end up in such a mess? Again there is more than one factor. The public reaction on the internal investigation report is an obvious one. Another is how the leaders in the first hectic days stood on the television screen and told everyone that Rørosbanen was safe. It had just proved not to be. People just don't buy that. I think we have all learned that honesty and a bit of humility are what are demanded in such a situation.

7 The Inspectorates' Position

The report of the independent governmental investigating commission stated that the Inspectorate in their opinion:

- Had too few employees.
- Had been too lenient with NSB and Jernbaneverket the first three years of its existence. This was particularly related to the philosophy of working at a systems level.
- Had been too weak when the departure procedure was being put into force without approval.

Their recommendation was that the Inspectorate should have more staff and perhaps report to another Ministry. More specific inspections and controls should be performed.

This was directly in opposition with the view Jernbaneverket and to some extent NSB had had on the Inspectorate in the time before the accident. This was well documented in earlier correspondence. The press put a lot of emphasis on this in their articles and in that way gave us a lot of support. The result is that the inspectorate's position has changed completely.

We are still small (14 employees) and we still report to the Ministry of Transport, but today we have the authority we previously only had in theory. That means we will not have a situation where changes are made without our approval. That means our workload in increasing. But that does not mean there are no more disagreements between us and Jernbaneverket and NSB. That's fine; disagreement brings discussion that brings us all further. But in the end we make the final decision.

This means of course also that we have to take a greater responsibility for the safety of the railways. If an accident like the one on Rørosbanen occurs again in, say, 10 years, we will not be let off as easily as we were this time. Authority means responsibility, we have to accept that.

8 The Track Ahead

To complete the picture a couple of events must be mentioned. The police also had an investigation into the accident. They reached a very similar conclusion to that of the commission, and the result was a fine of about £ 800.000 pounds given to Jernbaneverket. After some reflections Jernbaneverket accepted the fine. This is the biggest fine ever given a company after an accident in Norway. In connection with the revising of the national budget, Jernbaneverket asked for £ 23.mill. to pay for all the safety projects. They got £ 1,6 mill. Half of it had to be paid back as the £ 800 000 fine.

At the time of writing (September 2001) most of the actions following the report has been completed, but some are delayed. This was only to be expected with the tight scheduling. The reengineering of the interlocking is the furthest behind, and will not be completed until next spring. The task involved more work than expected.

The organisation of Jernbaneverket has undergone a lot of change. The most important change is that every region has employed a safety manager who reports directly to the director of the region. There is now a safety management system. Procedures have been produced and the safety handbook has been developed. The number of safety staff has increased. No changes are to be made without risk-analysis. The departure procedure is analysed and changes have been made to it.

But there is still a lot of work to do. There are still lines without a communication system. There are still lines without ATS / ATC (manual lines). These are very costly systems and developing them will take time and money. And are the public willing to pay for them? That remains to be seen!

The railway organisations are hard to turn. After being told over and over how important it is to keep the trains running, and to schedule, it is hard for staff on every level to suddenly shift their focus. Their manager might not be so very clear on the matter either. And the public still wants the trains to be on time.

But hopefully the railways in Norway are on the right track now. It might not be statistically significant, but so far this year the accident statistics have been good. The next step is to re-establish trust. That might still take some time.

Acknowledgements

I would like to thank John Tillotson and Øystein Skogstad for their invaluable help and support.

References

[NOU 2000:30] Norges offentlige utredninger 2000:30 Åsta-ulykken, 4. januar 2000, Statens forvaltningstjeneste, Informasjonsavdelingen, 2000

Reasons for the Failure of Incident Reporting in the Healthcare and Rail Industries

C.W. Johnson

Department of Computing Science, University of Glasgow,
Glasgow, Scotland, G12 9QQ.
johnson@dcs.gla.ac.uk
http://www.dcs.gla.ac.uk/~johnson

Abstract. Incident reporting systems have recently been established across the UK rail and healthcare industries. These initiatives have built on the perceived success of reporting systems within aviation. There is, however, a danger that the proponents of these schemes have significantly over-estimated the impact that they can have upon the operation of complex, safety-critical systems. This paper, therefore, provides a brief overview of the problems that limit the utility of incident reporting in the rail and healthcare industries.

1 Introduction

On the 30th November 1999, the Deputy Prime Minister, John Prescott, announced that the Confidential Incident Reporting and Analysis System (CIRAS) would be extended from the Scottish railway system to cover the entire network [4]. On the 13th June 2000, the Health Secretary, Alan Milburn, and the Chief Medical Officer for England, Liam Donaldson, announced the establishment of a centralised reporting facility for adverse incidents across the UK National Health Service (NHS) [5]. The Chief Medical Officer said; "At the moment there is no way of knowing whether the lessons learned from an incident in one part of the NHS are properly shared with the whole health service". The Health Secretary said; "Patients, staff and the public have the right to expect the NHS to learn from its mistakes so we can ensure the alarm bells ring when there are genuine concerns so they can be nipped in the bud". The more detailed statements that followed these media announcements made a number of claims about the benefits of incident reporting. These can be summarised as follows:

1. Incident reports help to find out why accidents do not occur. Many incident reporting forms identify the barriers that prevent adverse situations from developing into a major accident. These insights are very important. They help analysts to identify where additional support is required in order to guarantee the future benefits of those safeguards.
2. The higher frequency of incidents permits quantitative analysis. It can be argued that many accidents stem from atypical situations. They, therefore, provide relatively little information about the nature of future failures. In

contrast, the higher frequency of incidents provides greater insights into the relative proportions of particular classes of human 'error', systems failure, regulatory weakness etc.

3. They provide a reminder of hazards. Incident reports provide a means of monitoring potential problems as they recur during the lifetime of an application. The documentation of these problems increases the likelihood that recurrent failures will be noticed and acted upon.

4. Feedback keeps staff 'in the loop'. Incident reporting schemes provide a means of encouraging staff participation in safety improvement. In a well-run system, they can see that their concerns are treated seriously and are acted upon by the organisation. Greater insight into national and global safety issues can be gained.

5. Data (and lessons) can be shared. Incident reporting systems provide the raw data for comparisons both within and between industries. If common causes of incidents can be observed then, it is argued, common solutions can be found. However, in practice, the lack of national and international standards for incident reporting prevents designers and managers from gaining a clear view of the relative priorities of such safety improvements.

6. Incident reporting schemes are cheaper than the costs of an accident. These is an argument that the relatively low costs of managing an incident reporting scheme should be offset against the costs of failing to prevent an accident . This is a persuasive argument. However, there is also a concern that punitive damages may be levied if an organisation fails to act upon the causes of an incident that subsequently contribute towards an accident.

Numerous studies have described tools and techniques that are intended to help realise these potential benefits of incident reporting systems [33, 51]. Very few papers analyse the reasons why some initiatives have failed to yield safety improvements [27]. This is a significant omission. Some reporting systems only elicit a very small number of contributions. Those submissions that are obtained may only come from particular sections of the workforce [7]. Internal pressures prevent safety managers from responding effectively to particular contributions [30]. The following pages, therefore, summarise the reasons for the failure of incident reporting. The analysis is illustrated by case studies that are predominantly drawn from the healthcare and rail industries. This choice is justified by recent public attention and by the different management structures that characterise these two domains. It is important to note, however, that many of these problems affect a range of other industries, including aviation [27].

2 Unrealistic Expectations

It can often be a surprise to learn of the sheer scale of adverse events and near-miss incidents that affect certain safety-critical applications. Vincent, Taylor-Adams and Stanhope observe that between 4-17% of patients in acute hospitals suffer from iatrogenic injury [54]. Observational studies have found that 45% of patients experienced some medical mismanagement and 17% suffered events that led to a longer hospital stay [1]. It has been estimated that approximately 850,000 adverse events occur within the NHS each year [43]. 'Signals Passed At Danger' or SPADS provide a further example. These occur when a train passes a signal showing 'danger' without authorisation. There were over fifty of these incidents on UK railways in July 2001 [21]. This was nineteen more than in July 2000 but three less than the average for this month over the last six years. Eleven trains ran past the signal by more than 200 yards. One SPAD led to a derailment and two to track damage. In eleven of these cases, the driver had passed a signal at danger before. Since Ladbroke Grove, formal inspections have been required for all SPADS. This has resulted in over 200 formal investigations by Her Majesty's Railways Inspectorate (HMRI). The nature of these incidents is such that there may also be a far larger number of near-misses. Drivers often manage to avoid passing the signal by rectifying a potential problem 'at the last minute' [10]. The proponents of voluntary incident reporting have argued that these near miss incidents provide valuable learning opportunities. They can yield significant insights because they tell us about potential protection mechanisms that might be used in similar circumstances. They also tell us about the potential vulnerability of the system to future incidents in which other drivers might not be so fortunate.

Many organisations have high hopes when they introduce voluntary incident reporting systems. These schemes are not only intended to improve safety by identifying the potential for future failure. It is also, often, hoped that they will reduce costs by avoiding the negative consequences of previous failures. This dual role is illustrated by the objectives that have recently been established for incident reporting within the NHS:

> "...the Department of Health should establish groups to work urgently to achieve four specific aims: by 2001, reduce to zero the number of patients dying or being paralysed by maladministered spinal injections (at least 13 such cases have occurred in the last 15 years); by 2005, reduce by 25% the number of instances of negligent harm in the field of obstetrics and gynaecology which result in litigation (currently these account for over 50% of the annual NHS litigation bill); by 2005, reduce by 40% the number of serious errors in the use of prescribed drugs (currently these account for 20% of all clinical negligence litigation); by 2005, reduce to zero the number of suicides by mental health inpatients as a result of hanging from non-collapsible bed or shower curtain rails on wards (currently hanging from these structures is the commonest method of suicide on mental health inpatient wards)." [43]

These are high expectations. Unfortunately, many existing reporting systems have not delivered safety improvements on the scale that some people have predicted. For example, previous paragraphs have described how John Prescott has promoted the expansion of the Scottish CIRAS reporting system to cover the entire rail network. In spite of the perceived success of the CIRAS system, it is hard to demonstrate that Scottish railways have a significantly better safety record than other areas of the network. In June 2001, ScotRail was one of ten train companies warned by the Railways Inspectorate that it had not done enough to combat the problem of Signals Passed At Danger.

High expectations must also be contrasted with the very prosaic problems that have limited the effectiveness of previous incident reporting systems. For example, it can be difficult to persuade people to contribute reports about near misses or adverse occurrences. The Royal College of Anaesthetist's recent pilot study concluded that the self-reporting of incidents retrieves only about 30% of incidents that can be detected by independent audit. Jha, Kuperman, Teich, Leape, Shea, Rittenberg, Burdick, Segerand, Vander Vliet and Bates confirm this result in their studies of base-line incident frequencies [26]. Their work has detected adverse drug events using three different techniques. Firstly, they have used voluntary incident reporting. Secondly, they have use a computer-based analysis of patient records. Finally, they have performed exhaustive manual comparisons of the same data. In one study, they focused on patients admitted to nine medical and surgical units in an eight-month period [26]. Both the automated system and the chart review strategies were independent and blind. The computer monitoring strategy identified 2,620 incidents. Only 275 were determined to be adverse drug events. The manual review found 398 adverse drug events. Voluntary reporting only detected 23.

3 Fear of Retribution

The Cullen report into the Ladbroke Grove rail accident supported the use of confidential incident reporting as a means of eliciting information about adverse occurrences and near miss events. It was argued that: "(such a system) undoubtedly enables near miss incidents to be reported and receive attention". However, it was also argued that "in the longer term the culture of the industry would be such as to make confidential reporting unnecessary" [10]. Cullen notes that the UK rail industry is some distance away from an ideal situation in which confidential reporting systems would be unnecessary. In most industries, however, employees have a fear of retribution from participation in incident reporting systems. In large-scale national and international systems, this has led to the development of elaborate legal safeguards to protect potential contributors [2]. Unfortunately, most reporting systems lack the funding and the necessary managerial support to provide this level of assurance.

The relatively low participation rates in many reporting systems can be explained by a fear of retribution. In extreme cases, as the Federal Railroad Administration (FRA) notes, employees may even neglect medical treatment rather than expose themselves to workplace harassment:

> "FRA has become increasingly aware that many railroad employees fail to disclose their injuries to the railroad or fail to accept reportable treatment from a physician because they wish to avoid potential harassment from management or possible discipline that is sometimes associated with the reporting of such injuries. FRA is also aware that in some

instances supervisory personnel and mid-level managers are urged to engage in practices which may undermine or circumvent the reporting of injuries and illnesses." [12]

In the medical domain, a number of high-profile cases have acted as a powerful disincentive to the participation in incident reporting systems. For example, the High Court recently intervened to recommend the reinstatement of a surgeon who had expressing worries about the success rate of a colleague in his hospital. The trust initially refused to comply with the Court of Appeal's finding [6]. This parallels the case of Stephen Bolsin who first uncovered an unusually high death rate among babies undergoing cardiac surgery at the Bristol Royal Infirmary [32]. He subsequently claimed that he was unable to continue working in the NHS as a result of his 'whistleblowing' and was forced to move to a hospital at Geelong, near Melbourne. These causes have resulted in the Public Interest Disclosure Act (1998), which allows whistleblowing staff who feel they have been victimised to take their employers to an industrial tribunal. There is no limit to the compensation that can be awarded and employees simply need an "honest and reasonable" suspicion that malpractice has occurred or is likely to occur. Such protection has, however, proven to be insufficient to persuade employees to contribute to many voluntary reporting systems. For example, the 2001 Royal College of Nursing congress explicitly backed a call for action to protect nurses who 'speak out'. One of the delegates argued that whistleblowing was often seen as 'grassing up' or betraying colleagues. Theoretically, such additional protection should not be necessary under the 1998 Act. Some of these concerns can be explained by the informal pressures to conform to the norms of a particular working group. They can also be explained by the practical problems of preserving anonymity within small teams. Given the limited numbers of staff who perform particular tasks on particular shifts, potential contributors can often be identified through a simple process of elimination.

4   Reporting Biases

The previous section has argued that a fear of persecution can prevent staff form participating in incident reporting systems. The success of such schemes in eliciting contributions can also depend upon a number of more complex factors. For example, nursing staff contributed about 90% of all of the reports that have been submitted in a local intensive care unit over the last decade. 621 reports were submitted by nurses compared with 77 reports by medical staff [7]. This suggests that the reporting system may tell us a great deal about the execution of medical procedures. It may, however, tell us relatively little about more complex problems in the planning, coordination and administration of treatment within a department. At one level, it can be argued that this imbalance is due to the reluctance of senior staff to participate in an incident reporting system. However, these figures must be interpreted with great caution. For instance, it is important to consider the total number of staff who might contribute to such a system. Usually the team consisted of three medical staff, one consultant, and up to eight nurses per shift. The larger number of reports contributed by nursing staff can also be explained in terms of the involvement in, or exposure to, the types of workplace incidents that were solicited under this particular scheme. Nursing staff had the most direct contact with the patients who remain the focus of the reporting system. Hence, it can be argued that they have a proportionately greater opportunity to witness adverse events [7].

Automated logging and tracking systems provide means of addressing the problems both of low contribution rates and of biased participation in a reporting system. The proponents of such systems often have an unfortunate way of advocating their introduction; "competent personnel love them, while incompetent personnel loathe them" [11]. Such assessments hide the difficulty and expense that is often involved in interpreting the data provided by such systems. There is also a concern that any data will be used to punish rather than support staff performance through additional training. These concerns have acted as powerful barriers against the introduction of monitoring equipment onto UK trains. Recommendation 9 of the HMRI report into the accident at Watford South Junction advocated the use of these systems to monitor driving technique. In 1999, however, less than 20% of trains carried this equipment [21]. More recently, the action plan to implement the recommendations of the Southall accident report included steps to extend both voluntary incident reporting systems and automated monitoring equipment [20]. This link between voluntary reporting systems and automated monitoring is instructive. The success of reporting systems in aviation is often explained in terms of the pilot's fear that any incident may have been observed and reported by their colleagues or by tracking equipment. Participation in reporting system often provides a limited degree of protection and support in any subsequent investigation. However, it is less clear whether such a joint approach might also be extended to healthcare applications. Some initial steps have been taken to use computer-based tools to automatically identify adverse occurrences [26]. Such techniques are inevitably complicated by the difficulty of judging the severity of the patient's condition prior to treatment.

5 Poor Investigatory Procedures

Even if a reporting system is successful in attracting a large number of submissions, further problems affect the way in which an incident or near-miss is investigated. Theoretical issues complicate the task of determining what should be considered in any investigation. For example, Mackie argues that any event will typically have a number of effects [38]. Any individual is likely only to observe a subset of those effects. This 'causal field' is determined by the individual's ability to observe those effects but also by their prior expectation of what those effects might be. An individual's interpretation of cause depends upon the subjective frame of reference determined by their causal field. Mackie's ideas have important consequences because they imply that an investigator's work may be influenced both by their observations of the effect of an incident and also by their expectations about what those effects will be. For example, if an investigator develops an initial view about the causes of an incident then they may restrict their view of the causal field only to those system behaviours that provide evidence about those causes.

Mackie also argues that many effects stem not simply from a single cause but from a combination of factors that are termed 'causal complexes'. These ideas are reflected in the UK Health and Safety Executive's guidance on the incident and accident analysis that support railway safety cases:

"There is much evidence that major accidents are seldom caused by the single direct action (or failure to act) by an individual. There may be many contributing factors that may not be geographically or managerially close to the accident or incident. There might also be environmental factors arising from or giving rise to physical or work-induced pressures. There is often evidence during an investigation that some of the contributory factors have been observed before in events that have been less serious. Accident and incident investigation procedures need to be sufficiently thorough and comprehensive to ensure that the deep-rooted underlying causes are clearly identified and that actions to rectify problems are carried through effectively." [22]

This quotation also hints at another factor that complicates incident investigations. Statisticians and philosophers, such as Hausman [19], have referred to 'causal asymmetry'. This embodies the idea that if we know the cause we can predict the likely consequences. However, if we only know the consequences then it is far harder to unambiguously identify a single cause. Typically, many different combinations of events might result in similar consequences. It is precisely this asymmetry that complicates the task of incident investigation and makes it imperative that individual intuition is supported by appropriate investigatory techniques.

A number of more prosaic problems affect the investigation of near-misses and adverse occurrences. There are often insufficient resources to perform a detailed study of the context in which an incident occurred. In local systems, this problem is mitigated by the participant and the safety managers working knowledge of the systems that are described in any contribution. It can, however, be difficult for these local systems to derive independent or expert advice, for example about human factors issues. There is, therefore, a tendency to blame incidents on inadequate attention or on poor staff performance rather looking at the underlying causes of human 'error'. For example, the reporting systems in a local hospital used acronyms to remind staff to perform particular actions. TAP stood for Tap Aligned Properly. Such advice provides short-term protection against certain classes of adverse events. However, their effectiveness declines rapidly over time. It can also be difficult to ensure that new staff are taught the various incantations that have been proposed. Subsequent study of many of the incidents that helped to generate these acronyms revealed that they were often 'work arounds' that were intended to support the use of poorly designed or faulty equipment.

Larger-scale reporting systems can avoid some of these problems by ensuring that their staff are trained in appropriate analytical techniques. Unfortunately, there is little agreement about which approaches might support the causal analysis of incidents in either the rail or the healthcare industries [34, 10]. This lack of consensus has important consequences. It can undermine confidence in the findings of any investigation, especially when there are misgivings about the intent or purpose of any enquiry. HMRI argue that "using the investigation of a SPAD as a means of determining whether, and if so what, disciplinary action should be taken, or as a means of determining questions of liability, for example as between companies, tends to discourage full root cause analysis" [21]

6   Flaws in the Systemic View of Failure

Criticisms of existing practices have prompted the re-training of investigators across the UK rail industry. This has advocated a more 'systemic' approach to incident investigation and looks beyond catalytic failures, often characterised by individual human errors, to examine more distal causes. These tend to stem from managerial or organisation failures [45]. This view has been embodied in HMRI guidance that rejects the identification of errors as root causes of incidents on UK railways:

> "In these criteria the term 'root causes' includes consideration of management' s real and perceived messages to workers, environmental and human factors, as we ll as plant failures and inadequate procedures. Human errors arising from poor operating conditions, procedures, management expectations or plant design are not root causes; the predisposing factors are." [22]

Unfortunately, a number of criticisms can be raised against the way in which this systemic view has been interpreted by many safety-critical organisations. Previous sections have argued that adverse occurrences and near miss incidents stem from causal complexes that are difficult to predict. The problems of causal asymmetry also make it difficult to be certain about the precise causes of any mishap. In consequence, many have argued that failures are 'emergent properties' that characterise complex, safety-critical systems [36]. Perrow, in particular, has argued that we may have to regard 'normal accidents' as the price for technological innovation [44]. A number of objections can be made against this view. The conditions for failure do not suddenly emerge from the application of new technology. Many individuals within safety-critical organisations are often well aware of the potential for an adverse occurrence [46, 53]. In contrast, management pressure to attain other commercial or organisational objectives act to stifle their concerns. Incident reporting can have a limited effect in reiterating the importance of safety concerns, that are typically already known to staff within the organisation. It does not, however, provide a panacea for the deeper organisational issues that have been identified by systemic views of failure.

There is also a deep irony in the systemic view of failure. It starts with the premise that incidents stem from managerial and organisational causes. These create the context in which individual human errors and systems failures can occur. Many managers have, however, argued that the difficulty of predicting these error-inducing contexts helps to absolve organisations from responsibility for particular failures. The implications of the systemic approach to failure can be seen in the words of Daniel Goldin; the head of NASA, when he spoke to the engineers and managers who had been involved in a series of unsuccessful missions to Mars: "As the head of NASA, I accept the responsibility, if anything, the system failed them" [40]. The tension between individual responsibility and the systemic causes of incidents is apparent in this citation. It is also apparent in the behaviour of senior managers within both the rail industry and the health service. At one level, they help to create and control the context in which adverse incidents occur. At another level, they cannot be expected to possess a detailed knowledge of the many different working practices that their staff adopt and that contribute to adverse occurrences. These observations help to explain the current popularity of incident reporting systems; they are seen as a means of communicating safety concerns to higher levels within the management of many organisations. It is, however, far harder to ensure that reporting systems provide accurate information about the potential risks that threaten safe and successful operation. Nor is there any guarantee that higher-levels of management will act on the information that they receive.

7 Analytical Bias

It is important not to underestimate the potential biases that influence the analysis of near misses and adverse occurrences. Over the past three years, we have conducted a series of interviews, surveys and observational studies of incident investigators and safety managers [49]. This work has helped to identify a range of influences that can affect the decision making processes that are intended to distinguish causal factors from the mass of other contextual information that is extracted from an initial report. The following list describes some of these biases. It is not intended to provide an exhaustive account:

Author bias. This arises when individuals are reluctant to accept the findings of any causal analysis that they have not themselves been involved in. For instance, a recent review by the FRA identified that incidents at US highway-rail crossings can trigger investigations by federal organisations, such as the NTSB and the FRA. They can also result in state level enquiries. In some states, responsibility is divided between public agencies and the railroad operators. Elsewhere, responsibility is assigned to regulatory agencies such as the Public Utility Commission, Public Service Commission, or State Corporation Commission. In other states, investigations involve representatives of state, county, and city jurisdictions. Both state and local law enforcement agencies will also be involved if an incident involves the enforcement of traffic laws. Local government bodies are given responsibility for operational matters related to crossings through their ordinances. The situation is slightly simpler for incident investigations in the UK. However, railway privatisation has created a situation in which conflict can arise between operating companies, Railtrack and the HMRI. This is neatly encapsulated in Anthony Scrivener's recent article on Ladbroke Grove entitled 'Pass the signal - pass the blame' [47].

Confirmation and Frequency Bias. Confirmation bias arises when investigators attempt to ensure that any causal analysis supports hypotheses that exist before an incident occurs. In other words, the analysis is simply conducted to confirm their initial ideas. Frequency bias occurs when investigators become familiar with particular causal factors because they are observed most often. Any subsequent incident is, therefore, likely to be classified according to one of these common categories irrespective of whether an incident is actually caused by those factors [25]. There are many examples of these two forms of bias in the handling of SPAD reports prior to the Ladbroke Grove accident. Cullen estimates that approximately 85% of all such incidents were classified as the result of driver 'error' [10]. The frequency of such findings helped to reinforce this analysis as an acceptable outcome for any SPAD investigation; "I am led to conclude that the ready acceptance of blame by drivers, encouraged by the no blame culture, may have contributed to this poor analysis of root causes". The subsequent report argued that operating companies should review their incident investigation practices to ensure that there is no presumption that driver error is the sole or principal cause of SPADs.

Recognition bias. This form of bias arises when investigators have a limited vocabulary of causal factors. They actively attempt to make any incident 'fit' with one of those factors irrespective of the complexity of the circumstances that characterise the incident. These pressures can be illustrated by the response to initial reports of problems in the performance of cardiac surgery at Bristol Infirmary. The Society of Cardiothoracic Surgeons of Great Britain and Ireland discussed the reports of poor outcomes in 1989. Further information emerged during site visits in 1990. The sub-optimal results were attributed to the low volume of work because an increasing number of cases was widely believed to be associated with better outcomes. The eventual enquiry argued that "the focus on throughput may with hindsight be thought to have distracted attention from further inquiry, as the Bristol results, with the exception of the figures for 1990, showed no real improvement" [32].

Political, Sponsor and Professional bias. Political bias arises when a judgement or hypothesis from a high status member commands influence because others respect that status rather than the value of the judgement itself. This can be paraphrased as 'pressure from above'. Sponsor bias occurs when a causal analysis indirectly affects the prosperity or reputation of the organisation that an investigator manages or is responsible for. This can be paraphrased as 'pressure from below'. Professional bias arises when an investigator's colleagues favour particular outcomes from a causal analysis. The investigator may find themselves excluded from professional society if the causal analysis does not sustain particular professional practices. This can be paraphrased as 'pressure from beside'.

The influence of these workplace issues can be difficult to assess. For example, the FRA Safety Board conducted an analysis of incidents from January 1990 to February 1999. This found that only 18 coded 'operator fell asleep' as a causal or contributing factor. The NTSB found these figures difficult to believe given the prevalence of such incidents in other modes of transportation [42]. Two NTSB investigations that had found fatigue as a causal factor were not coded in the FRA database as fatigue-related but as a failure to comply with signals. A number of influences might explain such different interpretations of the same incidents. For instance, the FRA plays a significant role in the promotion of the rail industry as well as in its regulation. The NTSB focuses more narrowly on the investigation of safety-related incidents. In consequences, the political, sponsor and professional influences that act on those organisations will be quite different.

8 Rhetorical Bias and the Problems of Counter-Factual Reasoning

Counter-factual reasoning lies at the heart of most incident investigations [13, 35]. This takes the general form that 'if a causal factor had not occurred then the incident also would not have taken place' [37]. If an incident would still have taken place whether or not a event had occurred then it cannot be thought of as causal factor. It is important to stress that counter-factual reasoning is not something new, unusual or surprising. It is often used informally by investigators without realising that this is what they are doing. For example, the US National Transportation Safety Board (NTSB) investigation used this form of reasoning to identify the causal factors in a recent crossing incident; "...had the FRA grade-crossing closure program been more successful in eliminating grade crossings, fewer grade-crossing accidents might have occurred" [41]. Counter-factual reasoning is also used at a more detailed level in the same report; "the train 102 engineer might have seen the long combination vehicle sooner and been able to stop the train in time to avoid the collision if the semitrailer involved had been equipped with retro-reflective tape". This argumentation style can also help to exclude 'causes' that did not contribute to the incident; " ..the structural elements of the Northern Indiana Commuter Transportation District railcar 11 collision post that failed were overwhelmed by the force of the collision, and the post could not have prevented penetration of the steel coil, given the train speed and the weight of the coil" [41].

It is difficult to underestimate the prevalence of counter-factual reasoning in the analysis of adverse occurrences and near-miss incidents. It forms a major component of the techniques advocated by NASA [39] and the US Safety Systems Society [50]. Counter-factual reasoning can, however, pose numerous problems for incident investigation. For example, how sure can we be that an incident would not have occurred if a causal factor had not been present? Causal asymmetries suggest that many different causal complexes will have the same outcome. For instance, there are few guarantees in the previous incident that the engineer would have been able to avoid the collision even if the semi-trailer had been equipped with reflective tape. The problems of inattention and fatigue in previous incidents have shown that such safeguards do not provide guaranteed protection against adverse occurrences. The complex issues surrounding counterfactual reasoning is a research area in its own right. Byrne and her colleagues have conducted a number of preliminary studies that investigate the particular effects that characterise individual reasoning with counterfactuals [8, 9]. This work argues that deductions from counterfactual conditionals differ systematically from factual conditionals and that, by extension, deductions from counterfactual disjunctions differ systematically from factual disjunctions. This is best explained by an example. If we argue that "the train 102 engineer might have seen the long combination vehicle sooner and been able to stop the train in time to avoid the collision if the semitrailer involved had been equipped with retroreflective tape" readers will infer that the semitrailer was not, in fact, equipped with retro-reflective tape. This counter-factual style of argument can have such a persuasive effect that readers overlook contradictory evidence elsewhere in a report [31]. There are more complex examples of the inferences that readers draw from counter-factual arguments. The statement that 'either the brakes were applied too late or the train was going to fast' is a factual disjunction. Byrne argues that such sentences encourage the reader to think about these possible events and decide which is the most likely. There is an implication that at least one of them took place. The statement that 'had the FRA grade-crossing closure program been more successful or the semitrailer been equipped with retro-reflective tape then the incident would have been avoided' is a counterfactual disjunction. Byrne argues that this use of the subjunctive mood not only communicates information about the possible outcome of the incident but also a presumption that neither of these events actually occurred.

This theoretical work has pragmatic implications for incident investigation. If factual disjunctions are used then care must be taken to ensure that one of the disjuncts has occurred. If counter-factual disjunctions are used then readers may assume that neither disjunct has occurred. The distinction between counterfactual and factual disjunctions forms part of a wider concern to ensure that analytical biases are not hidden through the inappropriate use of language in incident reports. For example, rhetorical devices known as tropes can be used to increase the impact and effectiveness of everyday prose. They can also be used to achieve particular effects on the readers of an incident report. The following paragraphs provide a brief introduction to the techniques that have been used within the rail and healthcare industries. A more sustained analysis is presented in [29].

Amplification involves the restatement of an idea or argument. It often also involves the introduction of additional details. For example, the US Food and Drugs Administration (FDA) recently described actions that were taken in response to incidents involving nutritional supplements:

"The recognition of a cluster of cases was the key to the detecting of Eosinophilia-Myalgia Syndrome (EMS). Interactions among various specialists, including a family physician, hematologist, rheumatologist, clinical immunologist and epidemiologists, was crucial to this process.

Both EMS's clinical seriousness, and uncertainties surrounding its etiology, indicate the need for health professionals to remain vigilant regarding adverse events possibly associated with the use of L-tryptophan-containing dietary supplements, and to report such events to MedWatch." [14]

The first paragraph notes the importance of detecting the initial cluster of incidents using a reporting system. The second paragraph amplifies this by noting the importance of the MedWatch reporting system as a means of detecting adverse events that might be associated with L-tryptophan in dietary supplements. This technique can have the effect of drawing the reader's attention to a particular concept or idea. The amplification not only introduces new facts but it also supports and reiterates the arguments that are introduced in previous sentences. This technique can create problems when the amplification of particular aspects of a previous assertion can detract from other arguments or items of information. It is, therefore, important to establish the credibility of both an initial assertion and the subsequent amplification. For instance, an article about the cluster in the Albuquerque Journal News on the 7th November helped to trigger the FDA release of the public advisory on the 11th November. The initial cluster was not triggered by submissions to the reporting system, as might have been inferred from the use of amplification in the previous example.

Anaphora uses repetition at the beginning of successive phrases, clauses or sentences. It can create an impression of climax in which the repetition leads to a particularly important insight or conclusion.

"In April 1990, two more cases of sudden death associated with the use of barium enema kits were reported. A 41 year-old female complained of nausea shortly after insertion and inflation of the tip/cuff assembly, went into cardiac arrest within 30 seconds and underwent unsuccessful resuscitation efforts. In the third case, a 72 year-old female had an immediate reaction after the tip portion of the tip/cuff assembly was inserted prior to introduction of the barium contrast agent, went into vascular collapse and died." [14]

This example illustrates the successive use of the phrase 'A XX-year-old female' to build up descriptions of similar incidents. The investigator uses each successive sentence to 'stack up' evidence in a manner that will eventually support their analysis of common causes. It is important to emphasise that such techniques are not of themselves either 'good' or 'bad'. Rhetorical devices can be used to convince us of well-justified conclusions or to support half-baked theories. It is important, however, to be sensitive to the effects that such techniques might have on the readers of an incident report. For instance, the previous citation can be interpreted to provide readers with a clear summary of the evidence that supports the investigators' conclusions. It can also be interpreted in a more negative light. The repetition of such phrases may create an impression of certainty about the similarities between incidents that might not be justified by the evidence.

Antithesis uses juxtaposition to contrasts two ideas or concepts. This can be illustrated by the use of the terms 'properly' and 'improperly' in the following report into a needle-stick incident:

> "The housekeeper reported that he had used extra force to push the lid down, because the container was overfilled... Another housekeeper stuck her finger while removing a full sharps container from a wall bracket. .. When used properly, sharps containers can prevent needle-stick injuries. When used improperly, they can create a serious hazard." [15]

Here the consequences of 'proper use' are contrasted with those of 'improper use'. This technique is important because readers may make a number of additional inferences based upon such constructions. In this context, it is tempting to infer that the preceding injuries were sustained as the result of improper use, although this is not explicitly stated. It is also important to consider that the author presents no evidence to support their assertions about the consequences of proper and improper use. This is important because the citation emphasizes that idea that violation of proper procedures will result in injury. It does not consider that poor design might have created the potential for such serious consequences.

Asyndeton omits conjunctions between words, phrases and clauses. This technique creates an impression of 'unpremeditated multiplicity' [18]. The reporter can think of so many elements in the list that they hardly have time to introduce explicit conjunctions. There is also a sense in which this technique builds to a particular conclusion. This is illustrated by the opening sentences of a report into medical error; "a 62-year-old man came to the Emergency Department complaining of mid-sternal pain, shortness of breath, diaphoresis, and nausea after shoveling snow" [16]. Asyndeton creates precise and concise summaries. It, therefore, offers considerable stylistic benefits to more verbose explanations. There are, however, dangers. For instance, the use of such enumerations can create an impression of completeness where none was intended. There may have been other indications about the patient's condition that were omitted from the list. There is, however, a strong tendency for readers to regard the enumeration as complete unless the final conjunct is omitted; 'a 62-year-old man came to the Emergency Department complaining of mid-sternal pain, shortness of breath, diaphoresis, nausea after shoveling snow'. This creates a converse problem for the reader. The implied omission of closing conjuncts, as in the previous example, can lead to uncertainty about the information that might have been omitted from the list.

Conduplicato relies upon the repetition of key words or phrases at, or very near the beginning, of subsequent sentences. Conduplicato provides a focusing device because writers can use it to emphasise key features in preceding sentences. This helps to ensure that readers notice concepts or ideas that may have been overlooked when they read the initial sentence. This can be illustrated by the following quotation in which contributors stress the importance of "turning off" the ventilation sensors

"Medical intervention was needed to turn off the minute ventilation sensor in each pacemaker. When the sensors were turned off, the patients' heart rates returned to normal." [16]

Most contributors and safety managers draft incident reports without ever being aware that they are exploiting such rhetorical devices. They inadvertently construct prose that supports their arguments without explicitly considering the impact that their use of language will have upon their readers. They may inadvertently stress conclusions that are not well supported by the available evidence. They may also cast doubt on other findings that contradict their version of events. Unfortunately, this inadvertent use of rhetorical devices is often exposed at litigation. This should not be surprising. Many law courses, especially in the United States, include training in the use of rhetoric.

9 Classification Problems

Many regulatory and investigatory organisations have begun to codify information about previous incidents. This has numerous benefits. Firstly, the use of such codes can help to strip out the rhetorical effects that bias the interpretation of natural language accounts. Secondly, classification schemes provide key terms that can be used to access incident data in large, national and international databases. Unfortunately, a number of problems affect the practical application of this approach. If the codification of incidents is performed centrally then it is important that staff understand enough about the context in which an incident occurs for them to ensure that the correct codes are assigned. Alternatively, if incidents are to be codified at a local level then it can be difficult to ensure that different safety managers assign the same codes to similar incidents. For example, the FDA describe a case study in which a violent patient in a wheelchair was suffocated through the use of a vest restraint that was too small. The risk manager, JC, proceeded as follows:

"She finds the list of event terms, which was detached from the rest of the coding manual... She muses: 'Mr. Dunbar had OBS which isn't listed in these codes; he had an amputation which is listed; he had diabetes which isn't listed; and he had hypertension which is listed'. JC promptly enters 1702 (amputation) and 1908 (hypertension) in the patient codes. She then finds the list for Device-Related Terms... She reviews the terms, decides there was nothing wrong with the wheelchair or the vest restraint, and leaves the device code area blank." [17]

The resulting classification of 1702 (amputation) and 1908 (hypertension) provided few insights into the nature of the incident. This classification is more misleading than corresponding prose accounts even considering the potential biases that can be introduced through rhetorical effects and poorly constructed counter-factual arguments. The problems of incident classification also affect the rail industry. SPAD investigation procedures require that each incident is assigned a causal category. A recent HMRI report described how:

"Inspectors found there were difficulties with two of these categories: 'misjudgement' and 'disregard'. Although miscategorisation is not thought to be widespread, examples were found of SPAD incidents which had been categorised 'disregard', which actually seemed to be instances of driver misjudgement. An example was seen in RTMZ where a Virgin Trains driver had appeared to make every effort to brake at Coventry signal CY37 in poor weather conditions, yet the incident was categorised as 'disregard' rather than 'misjudgement'. Inappropriate categorisation should be avoided, otherwise it will reduce the credibility of the SPAD incident statistics and affect the rating of specially monitored drivers." [21]

A number of problems remain to be addressed even if incidents can be 'correctly' classified according to appropriate taxonomies. The most important of these relates to the storing and retrieval of large collections of codified incident reports. At present, incident reporting systems rely upon relational database technology. Each incident is classified according to a number of pre-determined fields. This approach has a number of consequences. It can lead to an extremely static classification system. There is no automatic means of reclassifying thousands of previous incidents if changes are made to a taxonomy. Some reporting schemes now hold more than 500,000 reports [2]. The scale of such systems creates problems if we must reclassify historical data to reflect changes in the coding scheme. If such changes are not made then there is a danger that safety managers may fail to discern that recent incidents form part of a wider pattern, which is obscured by weaknesses in the previous classification scheme. This problem is particularly acute when taxonomies are extended to describe human behaviour, as in the previous citation. The field of human factors research has changed rapidly over the last decade with an increasing focus on group interaction. However, few of these changes have been reflected in incident reporting systems because of the costs associated with manually analysing and re-classifying existing records.

Further problems affect the use of relational databases. The theoretical underpinnings of these systems are often poorly understood by the people who must use them. Safety managers, therefore, often rely upon pre-formulated queries to sort, filter and combine incident data. These queries are pre-programmed by system administrators who typically have a clearer understanding of the semantics of the commands that are being issued to the incident database. A consequence of this is that safety managers are often not being provided with the information that they think they are requesting each time they issue a query. It can also be difficult for safety managers to formulate the queries that they really would like to ask of their system because they lack the necessary technical knowledge about the implementation and operation of relational database technology [28].

We have experimented with a number of alternative technologies that address the problems, described above. Probabilistic information retrieval and conversational case based reasoning systems enable users to search for data without forming complex, structured queries [28]. A further benefit is that web-based search techniques can help to automate the indexing of large collections of incident reports. This avoids the overheads associated with the manual classification that may be necessary when changes are made to the underlying models that structure relational databases. Unfortunately, probabilistic information retrieval and conversational case based reasoning tools suffer from other problems. In particular, it can be difficult to ensure that particular queries yield appropriate levels of precision and recall. A system can exhibit poor precision if it returns many incidents that the user does not believe are related to their query. The user must then manually filter the large number of incidents that the system considers to be a match. Conversely, poor recall occurs when a system fails to return an incident that the user believes is related to their query. This can prevent analysts from determining that an incident forms part of a wider pattern. These problems are compounded by the computational relationship between precision and recall. Systems that provide good recall are often imprecise. Conversely, systems that offer high degrees of precision will often exclude incidents that ought to have been returned as a potential match.

10 Reliance on Reminders

The success of any reporting system depends on the interventions that are triggered by the information that it provides. There are many notable success stories where agencies have responded in a prompt and effective manner [15, 16]. Equally, however, a number of limitations constrain the use of incident data to inform safety management. Incident reporting systems often yield few surprises. The organisations that establish and operate these systems already have a good idea of the safety issues that affect their working practices. This can be illustrated by pioneering research into incident reporting within NHS hospitals in the North-West of England [52]. One study focussed on 19 incidents that were reported over approximately one month to an Accident and Emergency Department. The particular analysis technique that was applied to this data yielded a total of 93 potential causes. 45% of these related to organisational issues while 41% were classified as 'direct' human causes. The organisational causes included the need to secure external services. In particular, incidents were often triggered or exacerbated by the need to secure beds for the patients in the Department. They also included a lack of senior staff during peak periods. Direct human causes included problems that new Senior House Officers experienced in interpreting X rays. They also stemmed from a culture of learning from mistakes and a reluctance to contact senior staff. A similar study was then conducted into incident reporting within an Anaesthesia Department. This yielded 15 incidents with 78 root causes. 27% of these were identified as organisational issues, 40% stemmed from direct human causes and 26% were related to technical issues. The incidents were argued to illustrate less effective protocols than had been established in the Accident and Emergency Department. Several incidents indicated confusion over which drugs to stock and when. The study also revealed design problems with particular devices and the inadequate training that some staff received before being required to operate new systems. It is ironic that most of these issues had already been identified as significant problems by safety-managers within the hospitals. Such concerns do not, however, secure the resources and wider organisational support that is necessary to address many of these issues. This problem can be illustrated by a recent conversation with an NHS trust manager. He argued that incident reporting saves money by avoiding litigation and by reducing the amount of time that a patient might otherwise have to spend in hospital. Incident reporting does not, however, generate the resources that are needed to invest in addressing safety-related problems.

The impact of these financial issues on the effectiveness of incident reporting systems cannot be underestimated. The most frequent remedial action in one intensive care unit within an NHS hospital was to disseminate staff reminder statements. In the period from August 1995 to November 1998, 82 'Remind Staff?' statements were issued out of a total of 111 recommendations [7]. The 29 other recommendations concerned the creation of new procedures or changes to existing protocols (e.g. 'produce guidelines for care of arterial lines - particularly for femoral artery lines post coiling'), or were equipment related (e.g. 'Obtain spare helium cylinder for aortic pump to be kept in ICU'). None of the recommendations addressed the organisational or managerial issues that have been identified as a potential target for incident reporting systems and which are the focus for systemic views of failure. Such issues were beyond the scope of the system. This is not an isolated example. Similar patterns can be identified within aviation reporting systems [27]. The reiteration of well-known safety recommendations raises fundamental concerns about the utility of incident reporting systems. Human factors research points to the dangers of any reliance on reminders. Unless people are continually reminded then they are likely to forget the importance of safety precautions over time [24].

11 Inadequate Risk Assessments

There are many reasons why the recommendations that are derived from incident reporting systems often rely upon 'short-term fixes' rather than addressing the underlying causes of incidents and accidents. As mentioned above, there is a perception that they are a cost-saving measure. Reporting systems are, therefore, isolated from the revenue streams that might otherwise support necessary investments. Incident reporting schemes are also poorly integrated into wider forms of risk assessment. I recently witnessed the bizarre situation in which a design team were using Bayesian techniques to derive best estimates for reliability data while others, in the same organisation, had numerical data for the same faults [30].

Some incident reporting systems do not conduct any formal risk assessment for the near misses and adverse occurrences that they identify. This creates problems for anyone who wants to exploit the incident data. There is no way of distinguishing whether one incident must be addressed before another even though one might have a relatively high probability of recurrence and the potential for severe adverse consequences [29]. In other organisations, groups have drawn attention to such high-risk incidents without provoking an appropriate response:

> "During the almost five years preceding the Ladbroke Grove accident, there had been at least three occasions when some form of risk assessment analysis on the signaling in the Ladbroke Grove area has been suggested or proposed. The requests were: the Head of Technical Division's letter of 11 November 1996 which requested a layout risk assessment of the re-signaling (paragraph 43); the Field Inspector's letter of 16 March 1998 to Railtrack (paragraph 64); and the Railtrack Formal Inquiry of 1 July 1998 (paragraph 66). In addition there was an earlier request for details of measures taken to reduce the level of SPADs in the area around SN109 recorded in the Head of Technical Division's letter of 1st March 1995 (paragraph 39). None of these requests appears to have been pursued effectively by HMRI." [23]

Even when risk assessments are performed, there can be biases that emerge in the criteria that are used. This is often unavoidable. For example, risk assessment typically involves some appraisal of the frequency and consequence of an event. However, with a near-miss incident one cannot assume that any future recurrence will have the same outcome. Many reporting systems, therefore, assume the 'plausible worse case scenario' is an approximation to the potential consequences of a failure. Some of the problems of assessing the consequences of adverse incidents can be identified in W.S. Atkins' recent report into the investigation of SPADs on UK railways:

> "The system for selection of incidents for full investigation is skewed towards shunting incidents, which are often of low consequence (both actual and potential). This arises from the simplicity of the severity category system used to provide an initial classification of the seriousness of incidents. In part the system classifies by length of overrun, an approach that we consider to be unhelpful because an incident at a shunt signal with zero overlap results in a high rating which is often out of proportion to its seriousness. A long overrun on plain line (maybe past the next signal at green) is often accorded undue seriousness. Disregard of a cautionary aspect, followed by a very short SPAD (with potentially serious consequences had the brake application occurred fractionally later) receives a disturbingly modest rating. A further weakness with the severity category system is that no distinction is drawn between derailments on running lines and contained derailments on trap points. This is distinctly unhelpful because in the latter case it is liable to result in inappropriate amounts of attention and effort being focussed on intrinsically low consequence events." [48]

This quotation illustrates an important barrier to the successful implementation of incident reporting systems. Unless they are supported by clear guidelines to help assess the potential frequency and consequence of any recurrence there is little likelihood that management will allocate sufficient resources to rectify major safety problems. As we have seen, consequence estimates are error-prone and difficult to validate. Previous sections have also described the problems that arise in deriving accurate estimates of incident frequencies from reporting systems. Many incidents are not reported because of a fear of retribution. Automated logging systems provide a greater assurance that potential failures will be detected, however, a recent NTSB symposium identified numerous instances in which these systems failed to provide reliable data [3].

12 Conclusion and Further Work

The last three years has seen a rapid growth in the number and scale of incident reporting systems. The Ladbroke Grove accident stimulated a range of initiatives in the UK rail industry, including John Prescott's expansion of the CIRAS reporting system [10]. The Bristol Infirmary enquiry has had a similar impact on the UK healthcare industry [32]. Strong claims have been made about the potential benefits of these systems. Incident reporting applications are perceived to offer valuable insights into the near-miss incidents that have the potential to threaten future safety. They can also be used to elicit information about 'lessons learned' and act as an exchange for best practice [52].

This paper has, however, argued that significant barriers must be addressed before incident reporting systems can be successfully applied within many industries. These can be summarised as follows:

1. *unrealistic expectations.*
   Many people who initiate reporting systems expect reductions in the frequency and consequence of adverse events that are unreasonable given previous experience in running these schemes. These expectations are particularly problematic given that many types of incident will not be reported to confidential systems. There can be strong organisation and cultural barriers that prevent employees from disclosing information about their friends and colleagues;

2. *fear of retribution.*
   Some local systems enjoy good levels of participation while trusted individuals administer the scheme. Staff learn to trust the integrity of those individuals. However, when they are replaced participation rates may fall dramatically [7]. This effect is clearly linked to potential contributors' concerns that they will be viewed as 'whistleblowers' either by their colleagues or by those who administer the system.

3. *reporting biases.*
   Even once confidence has been established in a system, there are few guarantees that all staff will contribute incident reports. Variations in participation rates have been observed both within working groups at the same location, as in hospital systems, and between geographical regions, for example across

the rail network. Automated systems are increasingly being introduced to trigger investigations into near-miss incidents. However, some tasks cannot easily be instrumented. Many of the more specialised monitoring systems are unreliable and often provide 'false positives' that consume finite analytical resources. In consequence, it seems likely that reporting rates of less than 20-30% will be typical of many healthcare applications. These problems do not affect some reporting systems. SPAD reports provide a relatively accurate impression of the frequency of these events. However, the monitoring systems that help to detect these incidents tell us very little about incidents that *almost* resulted in a SPAD but that were narrowly averted by operator intervention.

4. *poor investigatory procedures.*
Once an adverse occurrence or near miss has been reported, it can be difficult to determine what factors should be included within an investigation. This is important for theoretical reasons because it can be difficult to identify salient factors within what Mackie terms the 'causal field' [38]. Hausman also points to the problems created by 'causal asymmetry' [19]. If we know the cause then we can determine the effects, however, if all we observe are the effects then it can be difficult to reach firm conclusions about the multiple possible causes of those effects. These theoretical problems are exacerbated by the resource constraints that affect incident reporting. Many organisations lack both the funding and the expertise to investigate more than a single causal hypothesis. This clearly limits the value of any insights that might be obtained from the analysis of near miss incidents.

5. *flawed systemic views of failure.*
The limited resources that are available to fund the analysis of many incidents are stretched by the recent emphasis on the systemic causes of failure. The proponents of this approach have urged investigators to look beyond the catalytic, triggering events of individual human error to look at the deeper systemic causes that are often related to organisational and managerial issues [36]. Many reporting systems avoid the practical and 'political' difficulties that such studies entail by limiting the scope of their analysis. Some of the local systems 'target the doable' [7]. There have also been examples of organisations whose upper levels of management have actively exploited arguments about the systemic causes of failure to mitigate managerial responsibility for particular failures. They have argued that the 'system' failed them in a manner that could not have been predicted before the failure. This stretches the interpretation of the systemic view of failure; it has also created a situation in which organisations apparently accept that it may not be possible to use any form of analysis, including incident reporting, to anticipate and respond to future failures.

6. *analytical bias.*
There are numerous forms of bias that can affect the analysis of incidents

once they have been reported. We have briefly described author bias, judgement and hindsight bias, confirmation and frequency bias, recognition bias, political, sponsor and professional bias. This is not an exhaustive list but it illustrates the difficulty of ensuring that any investigation is not hindered by 'undue' influences. These issues are particularly important in incident reporting when many stages of an initial investigation and analysis will be performed not by an external authority but by the organisation that was directly involved in the occurrence.

7. *rhetorical bias and the problems of counter-factual reasoning.*
A variety of rhetorical techniques can be used to 'hide' analytical bias within an incident report. It can be difficult to avoid using these tropes when constructing prose arguments to support particular findings. It is, therefore, important that both the readers and the writers of these reports are sensitive to the effects that these techniques can have. For instance, the repeated reference to particular items of evidence can indirectly increase the salience of that information. The psychological effects of rhetorical devices are mirrored by the unintended inferences that can be drawn from counter-factual reasoning. This style of argument takes the following form; 'X is a causal factor if the incident would not have occurred if X also had not occurred'. As mentioned, counter-factual reasoning can be fraught with dangers. There is often an implicit and unwarranted assumption that X did, indeed, occur [19].

8. *classification problems.*
Many organisations have responded to the problems of interpreting prose descriptions by adopting causal taxonomies. These initiatives form part of a wider attempt to classify incidents according to a range of different criteria. This offers numerous benefits. In particular, the elements of the classification be used as indexing terms in relational databases. Unfortunately, field studies have shown that few safety managers know how to use these tools to accurately extract information about previous incidents. Problems also arise when the items in a database have to be manually reclassified to reflect changes in a causal taxonomy. This can be particularly onerous for national systems that hold many hundreds of thousands of records. Several prototype systems have been developed to address these problems. For instance, we are using information retrieval techniques that were originally developed for mass-market web-based applications. These approaches are the subject of on-going research and currently suffer from poor precision and recall.

9. *reliance on reminders.*
Many reporting systems lack the financial resources that are necessary to address underlying system failures. These systems are, typically, seen as a form of cost reduction rather than as a form of income generation. This separation of reporting systems from sources of investment can result in recommendations that focus narrowly on 'quick fixes'. Studies of previous systems have seen a tendency to adopt a perfective approach in which operators are urged

to try harder to avoid future incidents. Such reminder statements provide dubious protection given that they must be continually reinforced if they are not to be forgotten.

10. *inadequate risk assessment.*
The design of safety-critical applications is typically guided by some form of risk assessment. Risk can be thought of as the product of the consequence and the likelihood of a particular failure. Incident reporting systems have been proposed as powerful means of informing risk assessments. They can provide quantitative data about the relative frequency of previous failures [34]. As we have seen, however, analytical and reporting biases undermine such statements. Similarly, the nature of 'near miss' incidents makes it very difficult to identify the 'plausible worst case scenario' that might inform any decision about the consequences of a future recurrence.

This is a partial list. For instance, we have not considered the powerful influence that a fear of media publicity can have upon the dissemination of safety-related information about previous mishaps. Similarly, previous paragraphs have not mentioned the conflicts that can arise when external incident reporting agencies must rely upon funding approval from the managers of the organisation that they collect reports about. A more complete introduction to the problems of incident reporting and a detailed explanation of potential solutions are presented in a forthcoming *Handbook of Incident Reporting* [29].

We have illustrated the problems that frustrate incident reporting using examples drawn from existing systems in the rail and healthcare industries. Many of these applications, especially within the healthcare industry, have been sponsored by individuals with a personal motivation for identifying safety issues in their workplace. Other systems operate on a far larger scale, such as the SPAD reporting process for UK rail operators. As we have seen, however, many of these diverse systems have faced remarkably similar problems. For instance, it is difficult to validate the findings of any causal analysis. It can also be difficult to assess the risks of future recurrences. It is regrettable that the proponents of recent initiatives to set-up national incident reporting systems have not taken more time to consider the range of technical problems that complicate the operation of these existing systems.

References

1. L.B. Andrews, C. Stocking, T. Krizek, L. Gottlieb, C. Krizek, and T. Vargish. An alternative strategy for studying adverse events in medical care. *Lancet*, (349):309–313, 1997.
2. Aviation Safety Reporting System. The Aviation Safety Reporting System. Technical report, NASA Ames Research Centre, California, United States of America, 2000. http://asrs.arc.nasa.gov.
3. K. Bolte, L. Jackson, V. Roberts, and S. McComb. Accident reconstruction/simulation with event recorders. In *International Symposium on Transporta-*

tion Recorders, pages 367–369. National Transportation Safety Board, Washington DC, USA, 1999. http://www.ntsb.gov/publictn/1999/rp9901.pdf.

4. British Broadcasting Corporation. Rail summit moves forward on safety. Technical report, News Staff, BBC, London, United Kingdom, 30th November 1999. http://news.bbc.co.uk/hi/english/uk/newsid_543000/543019.stm.

5. British Broadcasting Corporation. Plan to stop dangerous doctors. Technical report, News Staff, BBC, London, United Kingdom, 13 June 2000. http://news.bbc.co.uk/hi/english/health/newsid_788000/788805.stm.

6. British Broadcasting Corporation. Doctors back down in whistle-blower case. Technical report, News Staff, BBC, London, United Kingdom, 2001. http://news.bbc.co.uk/hi/english/health/newsid_1470000/1470590.stm.

7. D. K. Busse and D. J. Wright. Classification and analysis of incidents in complex, medical environments. *Topics in Health Information Management*, 20(4):1–11, 2000. Special Edition on Human Error and Clinical Systems.

8. R.M.J. Byrne and S.J. Handley. Reasoning strategies for suppositional deductions. *Cognition*, pages 1–49, 1997.

9. R.M.J. Byrne and A. Tasso. Deductive reasoning with factual, possible and counterfactual conditionals. *Memory and Cognition*, pages 726–740, 1999.

10. Cullen. *The Ladbroke Grove Rail Inquiry Part 1 Report.* HSE/Stationary Office, London, United Kingdom, 2001. http://www.hse.gov.uk/railway/paddrail/lgri1.pdf.

11. M. Durkin. Digital audio recorders: Life savers, educators and vindicators. In *International Symposium on Transportation Recorders*, pages 139–144. National Transportation Safety Board, Washington DC, USA, 1999.

12. Federal Railroad Administration. Fra guide for preparing accidents/incidents reports. Technical Report DOT/FRA/RRS-22 Effective: January 1997, Office of Safety, Federal Railroad Administration, Washington DC, United States of America, 1997. http://safetydata.fra.dot.gov/Objects/guide97.pdf.

13. T. S. Ferry. *Modern Accident Investigation and Analysis*. John Wiley and Sons Inc., London, 1988.

14. Food and Drug Administration. Clinical impact of adverse event reporting. Technical report, Department of Health and Human Services, Public Health Service, US Food and Drug Administration, Rockville, Maryland, USA, 1996. http://www.fda.gov/medwatch/articles/medcont/synopses.htm.

15. Food and Drug Administration: A. Morrison. Avoiding sticks from sharp containers. *User Facility Reporting Bulletins*, 1998. http://www.fda.gov/cdrh/fusenews/fuse25.pdf.

16. Food and Drug Administration: D. Dwyer. Sending the wrong signals. *User Facility Reporting Bulletins*, 2000. http://www.fda.gov/cdrh/fusenews/ufb33.html.

17. Food and Drug Administration: M. Weick-Brady. Those codes! *User Facility Reporting Bulletins*, 1996. http://www.fda.gov/cdrh/issue18.pdf.

18. R. Harris. A handbook of rhetorical devices. Technical report, SCC, Cosa Mesa, California, 1997. http://www.sccu.edu/faculty/R_Harris/rhetoric.htm.

19. D.M. Hausman. *Causal Asymmetries*. Cambridge University Press, Cambridge, U.K., 1998.

20. Health and Safety Commission. The Southall Rail Accident Inquiry Report: HSC action plan to implement recommendations. Technical report, Health and Safety Executive, London, United Kingdom, 2000. http://www.hse.gov.uk/hsc/south-01.htm.

21. Her Majesty's Railway Inspectorate. Report on the inspection carried out by hm railway inspectorate during 1998/99 of the management systems in the railway industry covering signals passed at danger. Technical report, Health and Safety Executive, London, United Kingdom, 1999. http://www.hse.gov.uk/railway/spad-01.htm.

22. Her Majesty's Railway Inspectorate. Assessment criteria for railway safety cases. Technical report, Health and Safety Executive, London, United Kingdom, 2000. http://www.hse.gov.uk/railway/criteria/index.htm.

23. Her Majesty's Railway Inspectorate. Internal inquiry report: Events leading up to the ladbroke grove rail accident on 5 october 1999. Technical report, Health and Safety Executive, London, United Kingdom, 2000. http://www.hse.gov.uk/railway/paddrail/inq-03.htm.

24. V.D. Hopkin. The impact of automation. In M.W. Smolensky and E.S. Stein, editors, *Human Factors in Air Traffic Control*, pages 391–419. Academic Press, London, United Kingdom, 1998.

25. D. Javaux. The cognitive complexity of pilot-mode interaction. In *HCI-Aero'98: Conference on Human-machine Interaction in Aeronautics*, Montreal, Canada, 1998.

26. A.K. Jha, G.J. Kuperman, J.M. Teich, L. Leape, B. Shea, E. Rittenberg, E. Burdick, D.L. Seger, M. Vander Vliet, and D.W. Bates. Identifying adverse drug events: development of a computer-based monitor and comparison with chart review and stimulated voluntary report. *Journal of the American Medical Informatics Association*, 5(3):305–314, 1998.

27. C.W. Johnson. Don't keep reminding me: The limitations of incident reporting. In K. Abbott, J.-J. Speyer, and G.Boy, editors, *HCI Aero 2000: International Conference on Human-Computer Interfaces in Aeronautics*, pages 17–22, Toulouse, France, 2000. Cepadues-Editions.

28. C.W. Johnson. Software support for incident reporting systems in safety-critical applications. In F. Koornneef and M. van der Meulen, editors, *Computer Safety, Reliability and Security: Proceedings of 19th International Conference SAFECOMP 2000*, LNCS 1943, pages 96–106. Springer Verlag, 2000.

29. C.W. Johnson. *A Handbook of Incident Reporting: A Guide to the Detection, Mitigation and Avoidance of Failure in Safety-Critical Systems*. Springer Verlag, UK, 2002 (in press).

30. C.W. Johnson, G. Le Galo, and M. Blaize. Guidelines for the development of occurrence reporting systems in european air traffic control. Technical report, European Organisation for Air Traffic Control (EUROCONTROL), Brussels, Belgium, 2000.

31. C.W. Johnson, J.C. McCarthy, and P.C. Wright. Using a formal language to support natural language in accident reports. *Ergonomics*, 38(6):1265–1283, 1995.

32. I. Kennedy. *Learning from Bristol: the report of the public inquiry into children's heart surgery at the Bristol Royal Infirmary 1984 -1995*. Command Paper: CM 5207. Her Majesty's Stationary Office, London, United Kingdom, 2001. http://www.bristol-inquiry.org.uk.

33. U. Kjellen. *Prevention of Accidents Through Experience Feedback*. Taylor and Francis, London, United Kingdom, 2000.

34. L. Kohn, J. Corrigan, and M. Donaldson. *To Err Is Human: Building a Safer Health System*. Institute of Medicine, National Academy Press, Washington DC, United States of America, 1999. Committee on Quality of Health Care in America.

35. P.B. Ladkin. Causal reasoning about accidents. In F. Koornneef and M. van der Meulen, editors, *SAFECOMP 2000*, Lecture Notes in Computing Science No. 1943, pages 344–355. Springer Verlag, Berlin, Germany, 2000.

36. N.G. Leveson. *Safeware: System Safety and Computers*. Addison Wesley, Reading, MA, United States of America, 1995.

37. D. Lewis. *Counterfactuals*. Oxford University Press, Oxford, UK, 1973.

38. J.L. Mackie. Causation and conditions. In E. Sosa, editor, *Causation and Conditions*. Oxford University Press, Oxford, 1975.

39. NASA. NASA procedures and guidelines for mishap reporting, investigating and record-keeping. Technical Report NASA PG 8621.1, Safety and Risk Management Division, NASA Headquarters, Washington DC, USA, 2001. http://www.hq.nasa.gov/office/codeq/doctree/safeheal.htm.

40. NASA (D. Goldin). When The Best Must Do Even Better" Remarks by NASA Administrator Daniel S. Goldin At the Jet Propulsion Laboratory Pasadena, CA March 29, 2000. Technical report, NASA Headquarters, Washington DC, USA, 2000. http://www.hq.nasa.gov/office/pao/ftp/Goldin/00text/jpl_remarks.txt.

41. National Transportation Safety Board. Railroad Accident Report Collision of Northern Indiana Commuter Transportation District Train 102 with a Tractor-Trailer Portage, Indiana June 18, 1998. Technical Report NTSB/RAR-99/03, NTSB, Washington, DC United States of America, 1999. http://www.ntsb.gov/Publictn/1999/RAR9903.pdf.

42. National Transportation Safety Board. Safety Study: Evaluation of U.S. Department of Transportation Efforts in the 1990s to Address Operator Fatigue. Technical Report Safety Report NTSB/SR-99/01 May 1999 PB99-917002 Notation 7155, NTSB, Washington, DC United States of America, 1999. http://www.ntsb.gov/Publictn/1999/SR9901.pdf.

43. NHS Expert Group on Learning from Adverse Events in the NHS. An organisation with a memory. Technical report, National Health Service, London, United Kingdom, 2000. www.doh.gov.uk/orgmemreport/index.htm.

44. C. Perrow. *Normal Accidents: Living with High-Risk Technologies*. Princeton University Press, Princeton, NJ, United States of America, 1999.

45. J. Reason. *Managing the Risks of Organizational Accidents*. Ashgate Publishing, Aldershot, UK, 1997.

46. S.D. Sagan. *The Limits of Safety: Organisations, Accidents and Nuclear Weapons*. Princeton University Press, Princeton, NJ, United States of America, 1993.

47. A. Scrivener. Special report on ladbroke grove: 'pass the signal - pass the blame'. *The Locomotive Journal*, pages 8–9, June 2000. Quoted extracts from evidence to Lord Cullen's inquiry into the Ladbroke Grove accident, http://www.aslef.org.uk/dox/loco_june_00.pdf.

48. G. Sitwell and S. Purcel. Assessment of Investigations into Signals Passed at Danger (SPADs). Technical Report BL2077 004 TR06, WS Atkins Rail Limited, under contract from the HSE, London, UK, 2001. http://www.hse.gov.uk/railway/spad/spadrep1.pdf.

49. P. Snowdon and C.W. Johnson. Results of a preliminary survey into the usability of accident and incident reports. In J. Noyes and M. Bransby, editors, *People in Control: An international conference on human interfaces in control rooms, cockpits and command centres*, pages 258–262, Savoy Place, London, United Kingdom, 1999. The Institute of Electrical Engineers. Bath, UK, 21-23 June 1999.

50. Systems Safety Society: New Mexico Chapter. System safety analysis handbook. Technical report, Systems Safety Society, Unionvile, VA, USA, 1997.

51. T.W. van der Schaaf, D.A. Lucas, and A.R. Hale. *Near Miss Reporting as a Safety Tool*. Butterworth-Heinemann, Oxford, United Kingdom, 1991.

52. W. van Vuuren. *Organisational Failure: An Exploratory Study in the Steel Industry and the Medical Domain.* PhD thesis, Institute for Business Engineering and Technology Application, Technical University of Eindhoven, Eindhoven, The Netherlands, 2000.

53. D. Vaughan. *The Challenger Launch Decision.* Chicago University Press, Chicago, United States of America, 1996.

54. C. Vincent, S. Taylor-Adams, and N. Stanhope. Framework for analysing risk and safety in clinical medicine. *British Medical Journal,* pages 1154–1157, 1998.

ISSUES OF LOW-SIL SYSTEMS

The Management of Complex, Safety-Related Information Systems

Ken Frith and Andy Lovering
Electronic Data Systems
Hook, Hampshire, United Kingdom

Abstract

Two (unrelated) articles in past editions of the SCSC newsletter dealt separately with safety of information systems and the safety analysis of complex systems. This paper returns to the joint themes from the management perspective, and discusses a number of problems that have been experienced when establishing safety management regimes for information-based, complex systems, and suggests some of the measures necessary to resolve these problems.

1   Introduction

In May 1998, Carl Sandom and Robert Macredie discussed [Sandom 1998] the specific dangers of ignoring the (functional) hazards associated with information systems. Prior to this, in January 1998 Benita Lawrence described [Lawrence 1998] the limitations that current safety assessment techniques have when dealing with complex, integrated systems, as these methods are largely designed to handle independent, deterministic systems. In this paper, we wish to focus less on the technical aspects of complex, safety related systems, and more on the problems generated for the provision of adequate safety management.

2   Background

Robust safety management is required for all safety related systems, this requirement being driven by the need to ensure their successful integration and delivery, and to maintain an acceptable level of safety through life until decommissioning. The safety management of complex systems, or more specifically systems that contain other systems, presents additional problems when trying to demonstrate that they are safe. Examples of such systems include:

- Air traffic control;
- Battlefield information systems;
- Railway signalling and communication systems;
- Hospitals;
- Naval platforms (and the maritime environment in general).

Despite the diversity of the above examples, they all share a common problem - how do the different technologies, sub-systems, equipments and humans combine for the overall system to be safe? This problem is further compounded by the inevitable use of sub-systems which, whilst they do not have any intrinsic hazards, nevertheless may contribute to the overall safety case (either positively or negatively). These include those sub-systems that we choose to call "information systems".

Although in EDS we work primarily in the communications and information systems domains, we are nevertheless required by customers to provide safety arguments for these products (whether produced by us or by another party). A question that we are frequently asked in this process, whether by our own colleagues, by customers, auditors or other contractors alike, is: "… but if all you are doing is transmitting and processing information, where is the hazard?" It seems that many programme managers producing information systems simply do not accept that their systems are safety related, and are therefore unwilling to accept the imposition of formal safety requirements on their systems.

The regrettable conclusion from this experience is that (despite much education to the contrary) too many engineers and programme managers still view their safety responsibilities from the narrow confines of their particular part of a project. The outcome of this is that resultant "safety cases" for some information systems examine only the intrinsic hazards of the system in question – that is, those hazards that would normally be associated with the hardware aspects of an information processing system, such as electrical and physical hazards. Whereas it is perfectly proper to conduct a risk assessment of these hazards, our concern really starts when it becomes clear that that is *all* that has been examined!

Essentially the safety programme proposed for a complex system must be capable of reducing the risk of the overall system to a tolerable level, and it should therefore be possible to structure it to support this aim. However, what is not always clear is how to achieve this, and this is due in part to a number of factors deriving from the ownership and management of the system and the complexity of its constituent parts.

So what harm can information do? Well, that all depends on the system it's used in, which leads naturally on to the subject of system complexity. This then gives the purpose of this paper, which aims to examine the twin problems of information systems and system complexity from a number of viewpoints, including the management of safety in such circumstances, the delegation of safety responsibility, the setting of safety targets and the involvement in the safety case of the various system, service and component providers.

3 Information Systems

Information processing systems are ubiquitous. Indeed, it can be claimed that any system that contains programmable electronics thereby contains a de facto information processing system. And as the SCSC is primarily concerned with systems containing programmable electronics, we are therefore intimately involved in the safety justification of such information systems. This may seem to be a simplistic assumption, but it does underpin many of the following arguments and thus should not be forgotten! However, for the purposes of clarification, the term "information system" is used in this context to cover all systems that either process incoming data into a required form for use by another system, store data for future use by other systems, or even those that communicate data between systems without altering it.

Now, clearly there are some instances where an information processing system actually does operate in isolation. The PC on your desk is arguably one such, in that it is used to import and process information, and then to output that processed information to another source, be it a printer, removable media or the internet. I grant that one could argue that this is safety related if (for example) you choose to use your PC to write a rude letter to your neighbour, who subsequently comes round and thumps you! But that only serves to demonstrate the point of this paper: it is the *use* of the processed information in a *wider* system that causes it to become safety-related.

So where do we go with this? Clearly the starting point must be the global safety-related system in which the information processor is used. This immediately takes the problem beyond the bounds of the producer of the information processing system, and raises his question "how much responsibility do I have for safety assurance of this wider system?"

4   Complex Systems

The concept of a complex system is actually tautological. Indeed, the Oxford English Dictionary definition of "system" is a "complex whole, set of connected things or parts ...", so we are dealing with complexity as soon as we start defining and describing system boundaries. For our safety purposes it is thus useful to begin at the basic level, considering a conceptual safety related system model, as shown in figure 1 below.

IEC 61508 [IEC 2000] gives good guidance on how to manage a closely-coupled system such as this, and it is possible to consider that information systems in this model may well constitute part of the Controlling Functions, operating on the hazardous Equipment Under Control (EUC), or possibly part of the Risk Mitigation Measures. However, this concept becomes more difficult to sustain as the system's complexity increases, and the controlling or protection functions become further removed from the EUC.

This is particularly noticeable where responsibility for the individual components described above (for both procurement and operation) is devolved to separate authorities. At first sight this would seem to be an unwise situation that is best avoided, but there are many systems where such diversification of the safety model is inescapable. The following examples illustrate some of these problems.

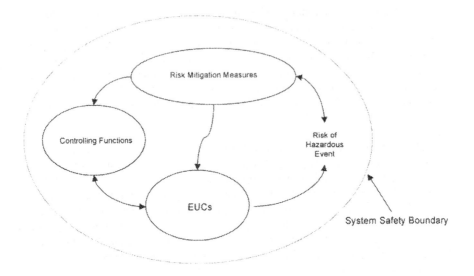

Figure 1 – Conceptual Safety Related System Model

4.1   Example 1: Battlefield System

A battlefield system-of-systems (super-system) is provided by a number of different projects, some of which are new procurements and some in-service systems (figure 2). The super-system consists of (essentially) sensor sub-systems (eg radar, optronics, air surveillance vehicles) feeding information processing sub-systems, in turn providing target data to delivery sub-systems (eg guns, missiles, aircraft etc).

These sub-systems are likely to be procured and operated by different authorities, the management problem being further compounded because the sensors, processing and delivery systems may not all be from the same service arm, coming from any combination of navy, army or air force (or even from different countries!).

At first sight, although this system is clearly complex, it can be argued that the super-system is conceptually straightforward, with an EUC (the delivery system) and a controlling function (the sensors and processing kernel). However:

- Who owns the overall safety case? The delivery system owners own the hazard(s) and may well be expected to provide the safety justification. However, the standard method of achieving this is to assume a given integrity requirement for the provided target data functions, and to build a safety argument based on their own closed-loop system. Bearing in mind the considerable number and diversity of delivery systems, all operating in differing and unpredictable environments, one can begin to see the management problem.   The provided target data can be sent with impeccable correctness and timeliness, but the receiving hazard owner has no control over the logic, method and integrity that were applied during its collection and processing. It therefore becomes unrealistic to expect any one of these hazard owners to make a complete safety case for the entire super-system.

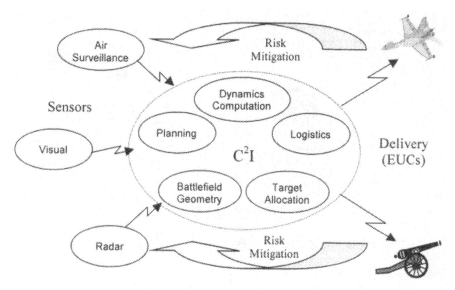

Figure 2 - Battlefield System Model

- Extending the IEC 61508 model, who provides the necessary protection mechanisms, if and when these have been identified? The delivery system owner can only achieve so much mitigation, and precious little when it comes to mitigating against the possible incorrectness of the provided targeting information. In such circumstances there has to be a superior system manager who has a more complete view of the super-system. In the absence of anyone else, this can put the onus on the managers of the information processing systems to provide mitigation – but is this correct?

- Management of change presents additional problems. Within this model, the introduction of a new delivery or sensor system should require a complete review of the information processor's Safety Case, and vice versa – again, is this a valid argument?

4.2   Example 2: Medical Diagnostic System

In an alternative, completely different context, consider a hypothetical medical diagnostic aid for a practitioner to provide logical diagnosis based on symptomatic, case history and epidemiological information (figure 3). Such expert systems are becoming more frequently available, and their safety is often justified by the argument that a skilled (human) practitioner is in the processing loop. Thus he/she is able to provide some form of mitigation against incorrect data or false machine diagnosis, by virtue of being able to make a reasoned (if judgmental) assessment of the risk involved in relying on the machine's advice.

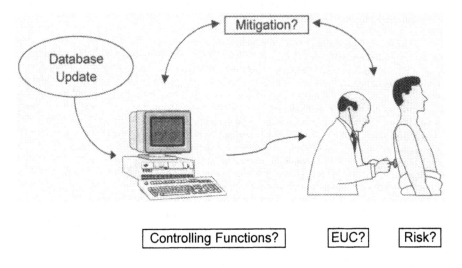

Figure 3 - Medical Diagnostic System Model

The whole (diagnostic aid, patient and doctor) is certainly a safety related information system, but is it complex? Considering that the patient and doctor interfaces are numerous and varied (medical conditions, practitioner skills etc), and that the database is necessarily fluid (having frequently updated knowledge, epidemiology and patient histories), then it almost certainly is complex. More importantly, it becomes complex by virtue of the split safety responsibilities: between the producer of the diagnostic aid, the data provider, the practitioner using it and the patients themselves.

This would not normally be a major problem, being simply another system with a human operator (and therefore subject to human error considerations), except that human factors analyses have shown that the use of such systems causes a dependency creep. The human increasingly relies on the machine analysis (particularly where this has a reliable history), and in consequence dilutes the mitigating analysis that would otherwise be applied. An example of this is the recently revealed case in a UK hospital of the millennium bug (failure to read the year 2000 correctly) having caused pregnant women's ages to be miscalculated, resulting in "low risk" assessments for Down's syndrome being given when "high risk" assessments were warranted . This problem often manifests itself when an apparently reliable system is changed − for example an updated database is supplied, or a lower-qualified practitioner is used, without the previous level of integrity checking and safety justification. Added to this, such systems may increasingly be used in areas which are more time-critical (for example in Accident and Emergency), where there is insufficient time for considered mitigation.

Although this second case is markedly different from the first, there is nevertheless the common theme of hazard ownership – ie who is liable if the practitioner uses (NB not "makes") a false diagnosis based on either incorrect logic or inaccurate data, neither of which he can control?

5   Hierarchical Systems

The examples above demonstrate that in many systems ownership is shared by more than one authority, which inevitably leads to confusion and conflict if not managed robustly. At the risk of introducing a dreadful oxymoron, let us consider a "simple" complex system. This takes the form of a super-system consisting of several sub-systems in some relationship to each other. Whereas this relationship may take a variety of functional, logical and physical forms, it is nevertheless possible to provide a hierarchical management structure that shows clear lines of procurement or operational responsibility, as follows (figure 4).

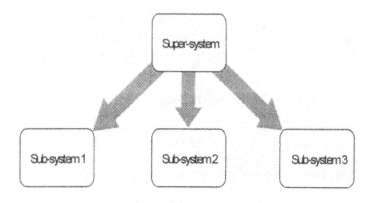

Figure 4 – Procurement or Operating Hierarchy

Achieving an adequate level of safety on such a procurement or operational model should begin by setting up an appropriate safety management structure. This is required initially to set and endorse the safety requirements on the system and its components, and then to facilitate the measures required to reduce the risk in each component (or to provide mitigation elsewhere if this cannot be done). Ideally the safety management structure should mirror the programme being implemented, as below (figure 5).

In this example, the programme requires a Safety Panel at the higher level to set the safety targets for each sub-system and manage the overall programme, and individual Safety Working Groups, which each reports to the Safety Panel. This reflects the aim of the safety programme, which is to demonstrate that the complete system will be safe. Thus the Safety Panel retains authority for the overall system safety and can impose this authority on the constituent sub-systems.

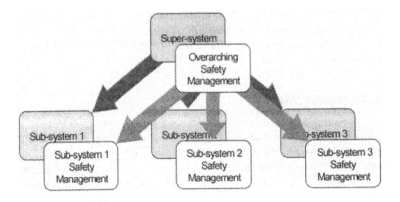

Figure 5 - Safety Management Hierarchy

However, whilst such arrangements are satisfactory for systems with clearly hierarchical procurement and operating relationships, many complex systems are not so easily defined. Although it is not our intention to address the functional and physical complexity issues in this paper, these nevertheless have a bearing on some of the problems of managing system complexity, and some of them are therefore discussed below.

6   Non-Hierarchical Structures

Not all programmes provide such a neat breakdown structure as described above, and may well consist of or incorporate more complex, non-hierarchical procurement or operating structures. Take, for example, controlling functions or protection mechanisms that rely not only on direct feedback from the EUC in order to decide and apply control or protection, but also on a variety of information sources concerning the systems background, environment, usage etc. In many cases, as Benita Lawrence points out in her work on safety cases for integrated systems [Lawrence 1998] the controls, protections or even the EUCs themselves may provide common failure points for numerous different systems, and the assumption that each sub-system can be treated independently becomes less tenable.

In such cases it may not be possible to impose a hierarchical safety management framework, and consideration of a separate, over-arching safety management organisation may be needed. This departure from hierarchical responsibility is a very serious problem in safety management. "Ownership" of the safety case (whether as developer, implementer or operator) is largely seen as an essential requirement of safety management, and this is seriously jeopardised when a system under your apparent management contains elements for which you have no responsibility, and over which you have no control. There have been several instances where this has created an unsafe situation, leading in many cases to actual accidents. The Herald of Free Enterprise accident was one such example, where conflict in management responsibilities resulted in unsafe practices and lack of adequate mitigation measures [Sheen 1987]. Indeed, this accident was influential in leading to the establishment by the International Maritime Organisation (IMO) of the International Safety Management (ISM) Code [IMO 1993], an internationally agreed code of practice for the management of shipping whereby clear safety management responsibilities must be defined.

7   Infrastructure Problems

Further to the above problem, many complex systems, including the examples described above, rely on components that can be regarded as "infrastructure". In this context we would define infrastructure components as those elements of the super-system that are not used exclusively for that system, and consequently would not be provided or managed by the super-system owner, or by anyone under his control. Such components may well include:

- Universal communications systems (including internet and other global communications);

- Generic databases and other information sources, where the content is not managed by the super-system owner (such as map and weather information systems in air traffic control (ATC));

- Other systems that contribute to the super-system in question, but are actually owned and managed by other authorities (eg the rail network and signalling systems for train operating companies).

Infrastructure components always cause problems to the designers and operators of safety related systems. One of the key problems is that the specification, design, implementation, management and operation of infrastructure elements are not controlled by the safety system owner. The consequence of this is that the system owner's safety justification must make safety assumptions about the infrastructure that can be supported throughout the life of his system. Essentially these assumptions fall into one of two alternative categories:

- The infrastructure has (and maintains) a demonstrable integrity. This sort of argument would be used by train operators when considering rail networks and signalling systems in their safety cases.

- The infrastructure cannot provide a satisfactory demonstration of integrity. Typically, non-dedicated communications and database systems would fall into this category. This does not mean that they cannot be used in safety related systems, nor does it imply that they are unreliable. It merely means that they cannot be shown to be dependable for high integrity functions.

This gives our system manager two options. In the first case, the infrastructure sub-system would be subject to the super-system's safety requirements, and would have to produce a supporting safety case, or otherwise contribute to the super-system's safety case. In the second case, the system manager has to assume that the infrastructure has no (or at least, limited) safety integrity, and will then have to include in the design sufficient mitigation to justify his own safety case. This mitigation could take several forms, for example:

- for communications infrastructures we can employ end-to-end error checking, for example by returning received data (or a measure of that data) to the originator for verification before use;

- for database information, the safety systems algorithms can use independent correlation (but beware of information loops) and gross error bounding.

However, whilst these are both achievable, real-time system performance issues (eg latency) may be difficult to resolve, and trade-offs between performance and safety will need to be examined closely and carefully justified.

8   Safety Requirements and Targets

The diversity in system criticality amongst the components of complex systems illustrates the problems that arise when defining the safety requirements for each sub-system. A communications service or information system is not directly hazardous, so initially one might assume that this system is not safety critical. However, a weapons targeting system most definitely is safety critical, and thus imposes requirements on the data it receives and passes via supporting information systems, that it is both accurate and timely. The result is that these supporting sub-systems are definitely safety related, and may indeed be critical. Therefore, setting the right safety requirements has to be managed using a top-down approach, as the indiscriminate setting of targets at the lower level cannot successfully be achieved. This is only possible by viewing the bigger picture; it is the safety of the entire operation that has to be demonstrated, via the contribution each system has to make in achieving this.

Thus for each sub-system (and on downwards in the system structure) the safety targets must be:

- *Meaningful.* It is pointless to pass on overall safety targets that have no bearing on the sub-system in question. For example, a target of 1×10^{-6} failures per operating hour is meaningless to a demand system (which category may well be considered to include some information systems).

- *Achievable.* It is regrettably common to see requirements placed on sub-systems that are evidently ludicrous. Even some reliability requirements imposed on high integrity software have been shown to be difficult to achieve (or even infeasible) [Johnson 1997], and safety cases based on such assumptions will clearly be flawed.

- *Verifiable.* As a corollary to these two requirements, the sub-systems' claims for meeting the given targets must be capable of being demonstrated for the safety case to be sustained.

Setting these targets will thus be a key factor in achieving a satisfactory, demonstrable safety case, and may necessitate the customer, procurer and designer rethinking how they align the programme management and organisation to facilitate the control and movement of safety requirements and targets.

9   Delegation of Safety

Assuming we can produce a management model for handling safety in our complex system, we must then delegate matters of safety to the stakeholders in the system. This is very easy to do on paper, and particularly so for a hierarchical system - for example a safety policy or plan can require that all parties abide by the plan and meet the set safety targets. But is this achievable in the real world?

As system complexity increases, the associated increase in management complexity often results in the consequence that contracted-out sub-systems can become divorced from the central system design, and thus lose visibility and control of the rationale and logic of the central safety argument. This can be a very real and serious problem; such isolation can often result in sub-system project managers being unable to understand, handle and meet the safety requirements and targets that have been placed on them.

In theory we can develop a system model that describes the system to its smallest elements, but in reality this would be both difficult to realise and to manage. Leaving aside the other aspects of project management, in safety terms the system designers (and ultimately operators) must ask themselves "how far down the chain do we pass safety requirements?" For example, whereas the owner of a hazardous operation may require his information processing system to be designed, and to operate, to a certain safety integrity level, can that system owner pass the safety requirement further down to his sub-contractors and suppliers? The answer must be "No" - nobody would expect that (for example) Microsoft would be required to provide a safety argument for their Office products because somebody else uses them in safety-related systems![1]

The system supplier has to assemble the system from a number of sub-systems, themselves comprising various elements each consisting of numerous components (the impression here of exponential fractal growth is deliberate).   These miscellaneous constituents may be widely sourced; some (eg software applications components) may be bespoke; others (eg hardware items) are almost certainly going to be commercially produced from stock items (Commercial Off The Shelf - COTS).   Is a system designer *really* going to approach each component manufacturer and insist that they achieve the overall system safety integrity levels (SILs)? Clearly there are cases when the component supplier may well be involved in the safety justification argument, but we would argue that in most cases, even at

---

[1] In fact, the reverse is becoming increasingly common, as suppliers put safety disclaimers on their products.

relatively high levels of the complex super-system, this is not practical[2]. For example, if you choose (perversely, perhaps) to use the internet as the communications medium within a distributed safety-related system, it would not be realistic to approach the ISPs (internet service providers) for assurances that they would meet your safety targets. Indeed, it would be unrealistic to assume that the precise nature or integrity of this particular communications medium could ever be fully defined!

What then is the answer? In our experience the pragmatic answer is that below a certain level it is better to make *no* safety assumptions about such components, and to expect no involvement of such component manufacturers in the safety argument. Just because we use Microsoft's software in a safety related system does not mean that we would expect them to be involved in providing safety evidence for our safety case. Similarly, we would probably not expect certain of our infrastructure providers even to be safety literate with respect to their products and services. It may well be valid to assume reliability or availability characteristics based on suppliers' provided (and demonstrated) data, but we have experienced problems in the past attempting to extract this information from COTS suppliers, and are probably not alone in this. However, even if satisfactory data is obtained, the onus for proving that these assumptions are valid, and that they support the safety arguments, lies with the safety system owner (perhaps even the super-system owner), and not with the component supplier.

10 Management of Safety in Non-Hierarchical Systems

How then can non-hierarchical systems, with all the above complexities, be managed? In some cases, the EUC owner (as the sub-system owner responsible for introducing hazards into the system) will be the appropriate manager for preparing the safety case (the ISM Code enforces this for shipping). However, it is more probable that the EUC owner will not have sufficient visibility or control of the super-system for this task. This is clear in many examples of complex systems. For example, the owner of a delivery system in the battlefield (eg a missile battery) may provide a safety case for his particular sub-system, but this can only cover risks that are visible to him, and cannot provide assurance of (for example) correct usage or targeting of that system. His system can perform exactly as designed and intended, but if somebody else's system misidentifies the target, the overall result will be unsafe. Similarly, one train operator may be acting perfectly within his safety case, but an accident occurs because of another operator's failings (such was the case at Ladbroke Grove [HSE 2000]). As the means to mitigate such accidents do not lie within the EUC owner's remit, he is unable to provide a satisfactory overall safety case.

---

[2] We are aware of the initiatives within the CASS scheme for certifying generic products for use in safety-related systems. Whilst they may well provide some answers to this problem, as these initiatives are still being developed and there are several aspects that still need to be addressed, no further coverage is intended in this paper.

However, some situations imply the need for an over-arching safety management responsibility, and indeed in the case of certain systems (rail, ATC) this belatedly seems to be the preferred route. So who provides this responsibility? We have the choice of an independent body being established, or of one of the constituent sub-system managers being given the task. Both have their merits and their faults:

- *Independent Regulation.* This has the advantage of independence, a broader super-system view, and a clear responsibility for safety. However, it also has the shortcomings (usually) of lack of authority and finance to research and implement safety measures and mitigations, and the lack of direct involvement in the system (hence lack of knowledge and understanding).

- *Internal Safety Management.* This has the advantage of involvement in (and therefore knowledge and understanding of) the system, and a greater ability to effect mitigation to some extent. However, the associated problem is the choice of sub-system manager best placed to provide an overall safety management system. This is often associated with a lack of finance and authority (ie the ability to impose the desired overall safety discipline), and the introduction of partisan attitudes (commercial and intellectual preferences).

This is where the manager of an information sub-system may be able to contribute. In many super-systems it is possible to identify an information sub-system that in effect acts as the IEC 61508 "controlling function". Such a sub-system necessarily has connectivity to most elements of the super-system, and its developers therefore require visibility and knowledge of the functional operation (and resulting hazard and accident sequences) of both the super-system and its constituent parts. The manager of such a key sub-system can thus be in an excellent position to provide effective safety management for the entire super-system, including responsibility for management and presentation of the overall safety case, both during development and in operation.

However, whichever Safety Management system is chosen, in order for it to be effective it *must* have the necessary responsibility and authority to achieve its function. This implies that it not only has powers of enforcement, but also has the funding to research, develop and implement required safety measures. This safety management authority will therefore be responsible for identifying safety requirement budgets, and for ensuring that these are ring-fenced within the super-systems constituent sub-systems.

11   Common Design Base

Crucial to the achievement of such a co-ordinated safety management structure, and to provide commonality across a complex system programme, it is evident that both the Safety Management organisation and all other stakeholders involved in the super-system safety case need a common visibility of the design. There are various methods available for achieving this, sometimes encompassed under the generic title "concurrent engineering". The Smart Procurement initiative adopted by the MOD's Defence Procurement Agency is such an example [AMS 2000], which requires the full involvement of all stakeholders in the programme, by the use of various concurrent engineering techniques including shared data environments. Without embarking on a treatise of the subject here, it is sufficient to say that these methods dispense with traditional sequential forms of systems engineering, utilising instead concurrent design methods that rely explicitly on full participation in all phases of the design by all stakeholders.

Such a project environment will naturally include all the information that supports the safety case, including the design description, safety management arrangements, hazard assessment, mitigation and the safety justification arguments. Sharing this will allow commonality of safety management plans and safety criteria (eg risk classification schemes) and ultimately lead to a common safety case.

12   In Conclusion

We hope that in the above discussion we have highlighted some of the problems that are experienced in a wide variety of information-based complex systems. Further, they demonstrate that even apparently simple safety-related systems can in fact prove to be disturbingly complex. Indeed, one could argue that all systems are in fact complex, and all contain information systems. This does not however help the designer a great deal, who has perforce to represent his system as a model suitable for analysis.

So is the problem technical or managerial? Well, probably both. There is clearly a technical dimension to the understanding of a system's entire complexity, and it is also clearly necessary to divide systems into understandable, manageable and procurable packages, while retaining a comprehensive overview of the super-system's overall functionality. However, there remains the problem of providing satisfactory safety management for such complexity. Straightforward systems with clear ownership can usually be managed under a hierarchical arrangement, with safety management paralleling the design and operational management. This is clearly the optimum management structure, if it can be achieved. However, where there is no obvious super-system manager, then a clear appointment of safety authority must be made. Whether this authority is to reside with one of the super-system's members (and who this is to be), or whether it is to be a separate entity, as some form of independent safety authority, is a decision that must be made as early as possible in the system lifecycle.

Whichever option is chosen, it is important to remember that Safety Management will not just happen, but has to be engineered to suit the system and all its stakeholders, and therefore put in place at the start of the programme (ie during concept), and maintained through life until disposal. This is not a new message, but it is surprising how often it is forgotten!

References

[AMS 2000]   The Ministry of Defence Acquisition Management System
                      (MOD AMS), v3.5, December 2000, Crown Copyright
                      1999-2000 (www.ams.mod.uk).

[HSE 2000]   HSE Internal Inquiry Report "Events leading up to the
                      Ladbroke Grove rail accident on October 1999", UK
                      Health and Safety Executive 5 April 2000.

[IEC 2000]   IEC 61508 Parts 1-7, International Electrotechnical
                      Commission 1999-2000.

[IMO 1993]   Annex to IMO Resolution A.741(18) adopted on 4
                      November 1993. "International Management Code for the
                      Safe Operation of Ships and for Pollution Prevention
                      (International Safety Management (ISM) Code)".

[Johnson 1997],   David M Johnson: "Increasing Software Integrity Using
                      Functionally Dissimilar Monitoring", Proceedings of the
                      Fifth Safety-Critical Systems Symposium, Springer-
                      Verlag 1997.

[Lawrence 1998]   Lawrence B: "Safety Cases for Integrated Systems",
                      SCSC Safety Systems January 1998.

[Sandom 1998]   Sandom C and Macredie R D: "Software Hazards and
                      Safety-Critical Information Systems", SCSC Safety
                      Systems May 1998.

[Sheen 1987]   Sheen, Mr Justice (1987): *"MV Herald of Free
                      Enterprise*. Report of Court No. 8074 Formal
                      Investigation", Department of Transport, London.

Engineering SCADA Products for Use in Safety-Related Systems

Brenton Atchison and Alena Griffiths

Invensys SCADA Development

brentona@foxboro.com.au, alenag@foxboro.com.au

Abstract

Supervisory Control and Data Acquisition (SCADA) systems are a class of control system used in a variety of application domains. Although SCADA systems are rarely relied on to provide the sole mitigation against high-risk hazards, they are frequently used to contribute to the management of hazardous situations, or to implement partial defences. As such, in some applications, SCADA systems are safety-related systems (as opposed to safety-critical systems). In determining the safety integrity requirements of a SCADA system, one must consider the environment in which the system is to be deployed, taking into account the availability of other hazard defence mechanisms and considering the proposed operational procedures. In this sense, the safety integrity requirements are generally not known until a safety analysis of a specific system in its target environment is performed. On the other hand, there is a growing demand for the use of standard, COTS, SCADA products that are combined to synthesize systems. This paper discusses the issues involved in engineering a base SCADA product for use in a diverse range of systems, both safety-related and non-safety-related. In particular, we address the issue of how to provide a base level of product assurance that can be used, if it ultimately proves necessary, to support system safety cases.

1   Introduction

Supervisory Control and Data Acquisition (SCADA) systems are a class of control systems used in a variety of applications to provide centralised control over a geographically diverse area. SCADA systems are generally produced in a COTS-like development model, with base products customised and configured for each customer.

Although they rarely directly control safety-critical functions, such systems may be relied on to assist in hazard management and are often vital to the management of critical civil infrastructure.

This paper discusses the development and assurance of products for SCADA systems. Section 2 provides an overview of SCADA systems and the products used to engineer them, and discusses SCADA system safety requirements in different application domains. Most companies who supply SCADA systems develop a number of SCADA products, which are synthesized as required to build systems. Section 3 explains why this product-based approach is unavoidable, but also points out assurance issues that arise when one seeks to develop a product for use in a diverse range of systems. Section 4 discusses an approach to the "product problem", which does not involve product certification.

2 SCADA Systems and Safety

This section provides an introduction to SCADA systems, outlines the products used to build such systems, and discusses the safety integrity requirements that can attach to SCADA systems in different application domains.

2.1 SCADA Systems – An Overview

SCADA systems are generally characterised on the basis of their architecture and control paradigm [Landman 2000]. A typical SCADA system architecture is illustrated in Figure 1. The rounded rectangles represent standard SCADA system components, the ovals represent optional, additional SCADA components often considered to form a part of modern SCADA systems. The rectangles represent external components with which SCADA systems sometimes interface.

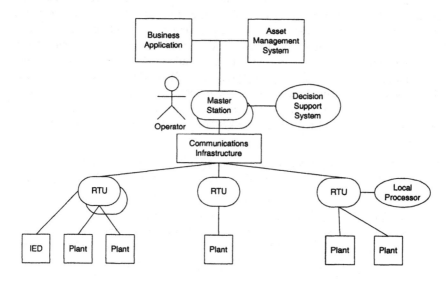

Figure 1 - Typical SCADA Architecture

In essence, a SCADA system consists of one or more *master stations* located in a control centre, with connections to many *remote terminal units* (RTUs) that are physically connected to plant. The SCADA system enables an operator at the control centre to monitor and control plant that is close to the RTUs. The master stations are sophisticated information systems, usually implemented on platforms such as Unix or NT, while the RTUs are simpler computing devices built on custom hardware, including custom I/O hardware.

RTUs can be spread over very large distances and are connected to the master stations by a communications infrastructure. Communications media can include dedicated wide-area-networks, radio links or public telecommunications networks. The physical communications media is typically not considered to be part of a SCADA system and a characteristic of SCADA is its ability to operate and recover in the context of unreliable communications.

Controls are issued from the master station and reach the plant via the connecting RTU. Information from the plant is telemetered back to the master station, again via the RTU. SCADA systems are characterised by an open-loop control paradigm, that is, control decisions are usually made by an operator based on the state of the plant, as indicated by displayed telemetered data. Operators are generally only present at the control centre, and one of the main functions the master station performs is to provide a suitable human machine interface that allows the operator to rapidly assess the overall state of the plant and so to make decisions about suitable controls to issue.

Modern SCADA systems often contain features that extend the control paradigm by providing local processing at the RTU and additional automation at the master station. Support for sophisticated local processing at the RTU can be useful to distribute processing load or to allow continued operation of plant during communications outages with the Master Station. Advanced master station applications such as decision support systems (DSS) provide advice to the operator about controls s/he should issue, based on a computer-based analysis of the state of the plant. DSSs are sometimes permitted to run in semi-automatic mode in which case controls are issued directly by the system, with the operator able to intervene if necessary.

There is also a growing trend toward the integration of business enterprise systems with SCADA systems. Examples of systems to which SCADA systems may interface include:

1) Asset management systems – information from the SCADA system can be used to provide all sorts of information to a system that optimises the owner's investment in their asset. Asset management systems can be used to do things like schedule preventative maintenance to optimise cost of ownership, tailor plant usage so that all plant ages gradually, etc.

2) Business applications – information from the SCADA system can be used to help owners make intelligent business decisions. For example, information from a power SCADA system may be used to drive a load forecasting application which permits owners to make best use of power purchasing contracts.

2.2   SCADA Products

Although every SCADA installation is unique, and systems differ substantially between application domains, there are nonetheless a number of common building blocks used in every SCADA system. For this reason, most companies that supply SCADA systems develop and maintain a number of SCADA products. The main products are the RTU and the master station. RTU and master station products are typically developed separately and may, in turn, be comprised of modules that can be acquired separately.

The base products can be configured in a large number of ways by the systems integration activity. Each application will use a different number of RTUs and master stations to match the distribution of their field equipment and requirements for load balancing and redundancy. Other configuration parameters include:

1) Selection of hardware modules to connect to field equipment and communications infrastructure

2) Mapping of field data and controls to a data model in the RTUs and master station

3) Definition of communications protocols and mapping of the data model to communications packets

4) Customisation of displays and the definition of alarm conditions

5) Definition of data archiving functions

6) Production of applications or RTU logic and integration with other information systems

2.3   SCADA System Safety Requirements

This section discusses the use of SCADA systems in some different application domains, and the safety integrity requirements that attach to such systems in these different domains.

Rail

In the Rail Industry, uses of SCADA systems include:

1) traction power distribution and control

2) monitoring and control of environmental plant on underground railways, (e.g. chillers, smoke extraction fans, tunnel ventilation fans, etc.)

3) an integration mechanism for many diverse systems (e.g. Passenger Address, Passenger Information Display, CCTV, miscellaneous station plant control such as escalators, lifts, lighting, etc.)

The main hazard scenarios associated with traction power distribution and control are electrocution of trackside maintenance workers, and electrocution of emergency services personnel during some sort of emergency in the vicinity of high-voltage power equipment (e.g. derailment or incursions onto the track). In such scenarios, the SCADA system is often deemed not to be safety-related, because work safety procedures involving physical isolation and earthing of HV equipment are relied on to protect trackside workers or emergency services personnel. Where such procedures are not used, the SCADA system must be used to isolate the relevant section and safety integrity requirements will probably attach.

In the case of environmental plant control, the SCADA system may be used to ventilate tunnels when a train is stranded or to extract smoke in the event of a fire. These functions generally give rise to safety integrity requirements. Sometimes such functions can be achieved via a station-based, local control panel, which means that the safety integrity requirements may be avoided or limited to the RTUs only.

Where SCADA systems are used to integrate other railway systems, the safety integrity requirements depend on the operational paradigm for the railway involved. While it is unusual for any particular function to give rise to a safety integrity requirement, it may be that the availability of an integrated picture of the current status of a range of railway subsystems is considered critical to the correct handling of emergency situations in general. As such, system availability becomes a safety integrity requirement.

In the rail industry, there is a general awareness of the potential safety implications of such systems. This is reflected in the fact that many contracts for the supply of SCADA systems will explicitly mandate a target safety integrity level for the system (SIL 1 or 2 is typical).

Oil and Gas

Typical functions of SCADA systems in the oil and gas industry include:

1) flow measurement and control

2) product quality monitoring

3) monitoring of pipeline control equipment, such as gas compressor and valves

4) integration with business systems to manage purchase and sales transactions

SCADA systems are generally not considered safety-related in the oil and gas industry since independent safety and emergency shutdown systems will manage pipeline hazards. Nevertheless, SCADA systems can provide an early indication of emerging hazardous situations, such as overpressure or pipeline leakage, and is often used as the first means of hazard control.

Triggering of an emergency pipeline shutdown is a costly exercise and can lead to other indirect hazards stemming from the loss of energy supply. The SCADA system also provides an essential service in the early diagnosis of pipeline equipment failures, for example failure of a key compressor, so is a critical factor in the overall pipeline availability.

Power Distribution

SCADA systems are used extensively to manage our electricity supply. Key functions include:

1) monitoring of substation current and voltage levels

2) monitoring of substation equipment and automatic recovery from some substation failure conditions

3) manual control of power supply

4) automatic regulation of transformers to achieve constant voltage under variable load conditions

5) isolation of substation equipment for maintenance

6) indication of personnel presence on site and alarming of emergency switch triggering

7) load measurement, trending and forecasting

Although a number of these functions are safety-related, SCADA is rarely used as the sole means for achieving them. For example, substation maintenance personnel are trained to physically isolate equipment before commencing maintenance. Despite not having direct responsibility for safety functions, it is possible that a SCADA fault or misuse, followed by a period of SCADA unavailability, could lead to insufficient power supply or loss of power.

Water Supply and Waste Water Treatment

Large towns and cities use SCADA systems to manage the supply of water and treatment of waste water. While the systems are different, many of the functions are similar and include:

1) measurement of water reservoir and waste water storage tank levels

2) remote water flow control, including overflow of waste water in storm conditions and supply of drinking water in high demand periods

3) isolation of valves for pipe maintenance

4) monitoring of pumping station equipment and power supplies

5) isolation of pumping station equipment for maintenance

6) monitoring of personnel presence in pumping stations

Water supply and waste water utilities can generally be managed through manual equipment control, indeed not all parts of a utility need be automated. In some instances, large pumping stations are augmented by local closed-loop control systems that monitor and manage water flow. The use of SCADA for large utilities allows for a centralised view of the utility and a more effective and cheaper management response.

Inappropriate operation of SCADA can lead to overflow of waste water in civil areas or to shortage of water supply. It can increase the likelihood of burst pipes by causing fluctuations in pipe pressure, or the stagnation of drinking water in reservoirs or pipes. SCADA is also essential in the detection of utility equipment failures and the subsequent coordination of a management response.

Summary

Across different industries we see a broad range of safety integrity requirements that may attach to SCADA systems. In most application domains, SCADA is not considered safety-related due to the presence of mechanical, procedural or independent electronic mitigations. Nevertheless, SCADA is often used as the initial detection and response mechanism for hazardous situations. Its advantage in providing centralised supervision and control of a complex system is important and unreliability or unavailability of the SCADA system may increase the overall risk of failed hazard management.

As our civil infrastructure grows more complex, SCADA systems are an integral factor in their effective management. As a result, system availability is often business critical, and system unavailability can have indirect safety implications. As such, many contracts for the supply of SCADA systems include clauses mandating quantitative reliability and availability targets.

Few application industries are as mature as the Rail Industry in terms of adopting a risk-based approach to safety integrity requirement determination [CENELEC 2001]. However, increasingly, contracts involve penalty clauses for failure to meet target availability levels. Given that SCADA systems are heavily software-based, the issues associated with attempting to demonstrate compliance with quantitative R&A targets can be similar to those faced in engineering software-based systems to an appropriate safety integrity level.

3   The Product Approach

As already mentioned, most companies that supply SCADA systems maintain a number of SCADA products. Individual systems are engineered using these products. This section explains why there is no practical alternative to the product-based development model, and then goes on to discuss assurance problems associated with this model. A number of possible solutions are considered, but each is shown to have one or more obstacles to overcome.

3.1   Why a Product Approach?

Despite discussion in some forums about assurance pitfalls that may be associated with using COTS and open systems to implement safety-related functions, there is nevertheless a seemingly irreversible trend towards the use of such systems. In the Rail Industry this trend is evident despite an awareness of the safety assurance issues. In industries that do not acknowledge any safety integrity requirements for SCADA systems, the use of COTS products and open systems/standards is standard practice.

The reasons why purchasers prefer COTS products are well-documented [Lindsay et al. 2000]. In brief, using COTS can reduce development risk and can increase reliability due to product maturity. It is very difficult for a solution provider seeking to develop a system "from scratch" to compete with a solution provider who offers a system composed of pre-existing components. The lack of competitiveness exists on several levels, including price, and market credibility.

There is another reason why purchasers of SCADA products tend to prefer COTS. Most purchasers have a significant investment in some sort of asset (e.g. power distribution plant). Such assets are maintained over a long period of time and, since the SCADA system assists in management of that asset, customers require a SCADA solution that will continue to be supported long-term. Over time, just as the purchaser will want to upgrade their plant, so they will look to upgrade their SCADA system. As such, purchasing a system whose component products have large user-bases means that, because a product with a large user base is likely to be upgraded, so it is likely the purchaser will be able to upgrade their SCADA system by purchasing component product upgrades. For this reason, most purchasers would prefer a product with a broad user-base, even if many of the installations are non-safety-related, than a product with a small user-base which, although it specifically targets the high-integrity market, is nevertheless at greater risk of being discontinued.

In practice, this means that most contracts for the supply of large SCADA systems are won by companies who possess (a suite of) SCADA products, that implement open standards communications protocols. The business and market forces are such that there is virtually no alternative to a product-based approach to the provision of SCADA systems.

3.2 Product Model Assurance Issues

If a SCADA system is determined to be a safety-related system, a safety case will need to be developed for that system. If the SCADA system is composed from a number of standard SCADA products, it will be incumbent on the supplier to show that the SCADA products are fit for the purpose for which they are intended. Typically, this will mean showing that a product correctly implements, to a suitable level of integrity, a number of functions that are safety-related in that system context. In this section, we consider a number of strategies that can be used to show that a component implements a safety requirement to a suitable integrity level. For each strategy, we explain why problems can arise in the context of the product model.

There are a number of COTS assurance strategies that can be employed [O'Halloran 1999]. We discuss how some of these strategies relate to SCADA.

High Quality Development Process

Many modern safety standards enable one to use evidence of a rigorous development process to support product integrity claims. Most companies developing SCADA products have generally developed them over a number of years and in response to an increasingly diverse set of requirements, including safety integrity requirements. During this period, the notion of best practice has evolved and development practices that were considered state-of-the-art 10 years ago would probably not produce evidence sufficient to support a modern safety case. As such, safety cases based exclusively on evidence associated with the product development process are difficult to support.

Operational Evidence

Claims of product integrity can sometimes be made given sufficient evidence of use in operation. However, most SCADA products evolve over time, with incremental releases, so it is difficult to build up sufficient operational time with each release. Furthermore, such products are configured differently in different systems and in any case exhibit a different operational profile so "proven-in-use" arguments are difficult to sustain. Some industries also display a parochialism that sees them reluctant to accept evidence of use in another industry as any guarantee for correct operation in the application industry.

Extensive Testing

Despite the bias in safety standards towards an assessment of the overall development process, many purchasers still consider evidence of extensive validation testing to be the best guarantee of correct performance. Extensive release testing is time-consuming, particularly because modern SCADA products can be extremely feature-rich. Moreover, SCADA products are also highly configurable, which adds another dimension to the test state space and compounds the problem further. This means that test-based assurance of correct operation of all product features in all system configurations will not be persuasive in most modern safety cases.

Safe Design Techniques

A common technique used when designing systems that perform many functions, only some of which are critical, is to partition the system so that critical functions are implemented in a safety-critical kernel, and the remaining functions are implemented elsewhere. Safety assurance for the non-critical functions is limited to showing that malfunction can not adversely impact correct and continuing operation of the critical functions. This design technique is not straightforward to apply in the case of a SCADA product, where the product must be suitable for use in a variety of systems, and where the sets of features considered safety-related in different systems can differ quite markedly.

Another common design-for-safety technique is the fail-safe approach. This approach requires that there exist a known safe state, and then involves engineering the system so that in the event of any anomaly, the system will fail in this safe state. It involves a sacrifice in the availability of some non-critical functions, in order to ensure that system reliability from a safety perspective is increased. Unfortunately, this strategy does not translate well to SCADA product engineering, since a state that is "safe" in one application may not be safe in another. To see this, consider the issuing of controls to plant. In rail traction power distribution, where the key hazard scenario is electrocution of staff working trackside, the safety requirement is that no spurious controls (say to reclose a circuit breaker) occur during the period when maintenance is in progress. As such, in this application, "no controls", whilst occasionally inconvenient, is nevertheless considered safe. Compare this with the environmental plant control application, where it is critical to be able to issue controls to smoke extraction fans when required. In such applications, the possibility of an occasional spurious control would be far preferable to the risk of non-availability of controls when required.

3.3 Possible Solutions to the Product Problem

Against the backdrop of a market preference for a product-based approach and the assurance problems associated with adopting that approach, the following generic options exist.

Multiple Products

One solution is to maintain multiple products, each targeted to the functional and assurance requirements of a particular market sector. Apart from compounding the cost of ownership, such a strategy is also unlikely to be welcomed by purchasers who want to buy a product with a large user base, so as to be assured of continued support for the product line.

Project-based Safety Certification

This solution involves developing individual safety cases for each deployed system. This approach has the advantage of relieving projects that have no safety requirements of any safety assurance overhead. On the other hand, duplication of effort is likely to result. Also, since the safety assurance evidence is developed by and stays with the project team, future product evolution may render this evidence irrelevant. As such, where future system upgrades involve product upgrades, and because the product upgrades may not have occurred with that particular system's assurance requirements in mind, safety case maintenance may become extremely expensive.

Product Certification

This solution involves achieving certification of a product to a certain safety integrity level. Typically, the certification is against a particular requirements baseline, and would involve the analysis, by an independent party, of the product development process and in-service performance. This approach has the benefit of significantly reducing the cost of producing system-specific safety cases (certification by a renowned third party is usually very persuasive). However, it also means that all product upgrades would need to be re-certified by the independent party, even those upgrades that were motivated by requirements for systems in non-safety-related domains. This places a considerable burden on the product release process, which may lead to product inertia. It also represents an overhead on the cost of the product as a whole, which may price the product out of certain markets.

4 Proposed Solution

This section proposes a solution to the problem of engineering a product that can be deployed across a diverse range of industries, with diverse functional and assurance requirements.

The basis of the solution is to distribute responsibility for the assurance of SCADA systems across an engineering organisation. Responsibility for the certification of systems lies with the market-specific, project engineering teams but is supported by evidence provided by the product development group. Requirements for evidence are forecast by market-specific, sales and marketing teams. The distribution of responsibility is illustrated in Figure 2.

Figure 2 - SCADA Assurance Model

The proposed assurance model allows each project engineering team to build a safety case in a form suitable to meet specific market and project demands, using product assurance as necessary. However, the model also requires the product team to develop and evolve the required product assurance. The model exploits the fact that the costs of engineering the required product assurance can be amortized not only across many system sales, but also across the entire life of the product.

The solution requires companies to embrace the idea that, for products of suitable size and complexity, a rigorous approach to development not only delivers product assurance collateral, it may also reduce product ownership costs (factoring in the cost of maintenance). Note that this idea is easier to embrace in a product context than in a single system context, where the contracts for initial supply and ongoing maintenance of the system are often let separately. Nevertheless, the idea is not universally accepted, and indeed many companies who maintain a safety-critical product consider product changes to involve enormous expense. In other words, they report that increased product assurance contributes to product inertia, rather than guards against it! We will return to this point later.

In line with the idea of distributed responsibility for assurance, our proposed solution entails a broader definition of "product", and hence a broader job description for product development groups. For computer-based systems, there is a tendency to think of a product as consisting solely of a physical hardware platform and the executable software that runs on it. We proposed broadening this concept so that a product is considered to consist of three baselines, as illustrated in Figure 3.

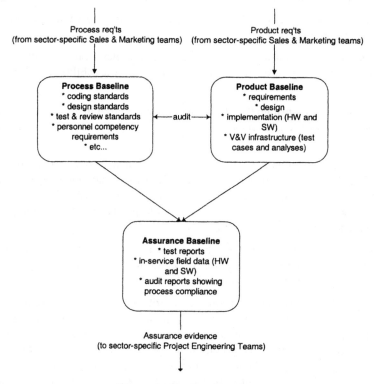

Figure 3 - Product baselines

The **product baseline** comprises the artefacts traditionally maintained and comprises:

1) the product requirements (both functional and non-functional). This could include specific process requirements demanded by some sectors (e.g. requirement to demonstrate 100% statement coverage at unit testing),

2) the product design,

3) product deliverables (i.e. hardward and executable software),

4) associated verification and validation infrastructure (e.g. test cases, test harnesses, etc.).

The **process baseline** describes how the product baseline was developed and includes:

1) the suite of meta-processes under which development of the product baseline occurs, such as the Software Quality Plan, the Verification & Validation Plan, the Configuration Management Plan, etc.; and,

2) the suite of specific processes used for various development activities, such as Coding Standards, Defect Reporting Guidelines, Change Management Guidelines, Technical Review Guidelines, etc.

The **assurance baseline** provides evidence of the integrity of the product and process baselines, including:

1) Evidence to demonstrate that product deliverables comply with product requirements (e.g. test results, design analyses, traceability matrices, user trials, etc.);

2) Evidence to demonstrate that the development occurred in accordance with the process baseline (e.g. documented evidence of reviews, checklists, and quality audit reports);

3) Data, both quantitative and qualitative, which reflects the in-service performance of the product deliverables. This would include a range of information, spanning issues such as observed in-service MTBF of a hardware component, through to the safety case used to justify inclusion of the product in (say) a railway traction control system.

The proposed solution also entails activities for the market specific groups whose job is to sell and engineer systems for a specific market. This engineering group must fulfill two key roles in this model.

Firstly, the sales and marketing team are contributors to the set of requirements contained in the product baseline. This will involve identification of trends in the market related to safety. This can range from physical requirements on the product deliverable (e.g. a compulsory requirement for redundant CPUs on the RTU), through to requirements on the product development process (e.g. a trend towards the use of formal source code analysis techniques as a compulsory pre-requisite for SIL 2 certification).

Secondly, the project engineering teams are required to make application-specific safety cases to get their systems accepted into service. The safety case is prepared in cooperation with the SCADA system procurer and end-users and will address aspects of the SCADA system, as well as its operational use. For this purpose, the project engineering teams are users of the artefacts that comprise the SCADA assurance baseline (in addition to being users of the product deliverables). Activities will typically include production of a hazard analysis to identify product safety requirements, with knowledge of the system architecture and operational context. A subsequent activity will then be performed to map the product, process and assurance baselines into a safety case that satisfies sector specific standards (e.g. establishing a compliance matrix from EN50128 tables to actual processes used during product development).

Our solution addresses product-based assurance issues by maintaining operational evidence for mature parts of the product, and process-based evidence for parts of the product subject to recent changes. Extensive validation testing is conducted, but is focused against a number of "standard system architectures", so as to reduce the possible configuration space needed for thorough testing. Those functional requirements specifically flagged by market-specific sales and marketing teams as being safety requirements are recognised as such throughout the design process. Safe design practices, in particular the use of a safety-critical kernel, are used where feasible. A critical aspect of the solution is to maintain the integrity of the product baselines as the product evolves. We specifically address this issue in the following section.

Dealing with change

For any solution to be workable, it must be possible to evolve the product at reasonable cost. Along with the expansion of the notion of "product" to accommodate three distinct baselines, so change management practices must be sophisticated enough to cope with changes in each baseline. This suggests the need for an approach to changes in product requirements and/or design, and to changes in the processes used to develop the product baseline. All changes must be handled in such a way as to preserve the quality of the evidence that is stored in the assurance baseline (e.g. poorly conceived or rushed changes can easily render pre-existing test data useless).

Of these two areas, change management for the processes used to develop the product baseline is an area that is not generally handled in texts on configuration management and change management. Careful thought must be given to managing the transition from the old to the new. In general, applying new process requirements to the existing product baseline would involve major re-engineering and as such would be practically unfeasible. It is therefore necessary to find a way to apply new processes to new developments and major product modifications, while permitting minor modifications or "bug fixes" to occur via the processes that led to the original development. This approach to the issue of process evolution is endorsed in some standards (e.g. EN50128 [CENELEC 2001]). However, this approach also leads to a situation where product assurance derives from different sources both across the product, and over time. This needs to be carefully explained and justified in safety cases using the product assurance baseline.

5 Conclusions

SCADA systems are used in many markets, but they are usually not thought of as safety-related systems. Section 2 presented an overview of SCADA systems and discussed the safety integrity requirements that may apply in different application domains. This analysis showed that while SCADA systems are never solely responsible for performing safety-critical functions, they occasionally perform safety-related functions (e.g. tunnel ventilation), and often contribute to early warning of and subsequent management of hazardous situations. Also, SCADA systems are relied on to manage the civil infrastructure used to ensure energy and water supply, and so prolonged system unavailability or faulty system behaviour can compromise those supplies, which can in turn have safety consequences.

SCADA systems are usually composed from a number of standard SCADA products. Indeed, as was discussed in Section 3, companies seeking to supply SCADA systems have virtually no commercially viable alternative but to maintain a suite of SCADA products. This creates difficulties from a safety assurance perspective, since the products need to be fit-for-purpose in the face of diverse functional and safety integrity requirements. There are many parallels between the issues pertaining to the use of COTS in safety-related systems, and to the problems faced by SCADA system and product suppliers.

A number of solutions to this problem were briefly considered but determined to be unsatisfactory for various reasons. Instead, in Section 4, we proposed an alternative that sees responsibility for assurance distributed across an organisation. In this model, the product group's responsibilities are expanded to involve maintenance of three baselines, and the application-specific project engineering groups use this information to build system and sector specific safety cases. An essential feature of this solution is an approach to change management that deals not only with changes in a product's functional requirements, but also with changes in a product's development processes. This feature is necessary to ensure that over time, assurance collateral is maintained and evolved, rather than replaced.

References

[CENELEC 2001]   CENELEC: European Standard ENV 50128, Railway Applications - Communications, Signalling and Processing Systems - Software for Railway Control and Protection Systems, 2001

[Landman 2000]   Landman, R. J:. "Supervisory Control and Data Acquisition Systems". Standard Handbook for Electrical Engineers. D. G. Fink and H. W. Beaty, McGraw-Hill: 10-147-10-168, 2000

[Lindsay et. al. 2000] Lindsay, P. and Smith, G.: "Safety Assurance of Commercial-Off-The-Shelf Software", in Proceedings of Fifth Australian Workshop on Safety Critical Systems and Software, Melbourne, Australia, Australian Computer Society, 2000

[O'Halloran 1999]   O'Halloran, C.: "Assessing Safety Critical COTS Systems." Journal of the System Safety Society 35(2), 1999

HUMAN FACTORS

Automating functions in multi-agent control systems: supporting the decision process

M. D. Harrison, P. D. Johnson and P.C. Wright
Department of Computer Science, University of York,
Heslington, York, YO10 5DD. UK
{mdh,pcw}@cs.york.ac.uk

Abstract

A challenge for designers of safety critical systems is to recognise that human issues are critical to their safe automation. Appropriate techniques for taking account of these issues should be integrated into the design process. This paper provides a brief introduction to a two-step decision procedure for deciding how to automate an interactive system. The method has received preliminary evaluation in aviation and naval contexts. It is intended for use early in the development of systems in larger scale collaborative settings with the aim of improving their safety and performance. The method aids the designer in identifying how to take advantage of the benefits offered by automation so that it does not interfere with the operator's ability to perform his or her job. It does this by guiding the matching of functions to a set of defined roles.

1   Introduction

For the automation of complex processes to be achievable with flexibility and safety, the need for procedures[1] that help decide how to automate becomes more important. Work systems are often complex interactive systems involving many people and many technology components. The work that is involved must be implemented in a way that is most compatible with roles that are designed for the people involved. Any automation should satisfy general criteria such as minimising workload, maximising awareness of what is going on, and reducing the number of errors. This paper is concerned with these procedures, known as function allocation, and as such is centrally concerned with the design of safety critical systems.

A number of methods of allocation of function have been developed since the early 1950s. In the main

---

[1] If they exist, see [Fuld 2000] for the case against existence.

- They fail to recognise fully the context in which the functions are to be automated. Functions are considered in isolation using general capability lists that describe what people are better at and machines are better at. More enlightened methods based around task analyses (for example MUSE, [Lim & Long 1994]) do not give explicit account of the environment in which the functions are performed. A richer description of the context for the analysis needs to be described.
- They are difficult to apply. They depend on an understanding of human factors that requires considerable training. For example the KOMPASS method [Grote et al. 2000] describes criteria for complementary system analysis and design such as process transparency, dynamic coupling, decision authority and flexibility. These criteria are described in terms that are very difficult to interpret by systems engineers.
- Methods do not use design representations that are familiar to systems engineers. Allocation decisions are usually binary. As a result a function is given the attribution "to be automated" or not and consequently must be described at a low level of granularity. Functions are not presented in a form that is easy to assimilate and use by software or systems engineers.
- They focus on the individual relationship between human and device and do not take into account the broader collaborative system and the roles defined therein.
- They offer no guidance about how dynamic allocation of function should be implemented. This problem becomes more relevant as the possibility of dynamic adaptation becomes more realistic.

More recently, methods of function allocation have emerged ([Older et al. 1997] and [Johnson et al. 2001b] for reviews) that take more account of some of these issues. Two methods: referred to herein as the 'Sheffield method', developed at the Institute of Work Psychology, University of Sheffield [Older et al. 1996] and the 'York method', developed at the Department of Computer Science, University of York [Dearden et al. 2000] overcome some of the deficiencies described above but not all. In particular, the Sheffield method is weak on use within a design context and the York method does not address the collaborative issue. What we describe here is a modest extension to the York method aimed at addressing some of the concerns about collaborative issues. Further extensions, not discussed here for lack of space, aim to integrate the inputs and outputs to the method with the notations of UML [Rumbaugh et al. 1999].

If function allocation were to be made an explicit step in the systems engineering development process, it would provide an ideal point to integrate human factors techniques into the development process. The desire for such integration exists because human factors techniques offer solutions to problems faced by engineers in the design of highly automated control systems. Designers of such systems face a dichotomy that can best be illustrated by way of an extreme design philosophy, that of automating everything possible. In a modern control system this could be achieved for a high proportion of the functions and has several advantages. Human operators are a source of error, and by minimising their role these risks can be reduced. Automation also offers the potential for greatly reducing costs. However,

Bainbridge [Bainbridge 1987] notes that this design philosophy is flawed because humans cannot be eliminated entirely. When situations arise that the automation cannot handle, the operators are expected to step in and resolve the situation. Their ability is impaired by the high levels of automation that have been introduced, resulting in out-of-the-loop performance decrements and loss of situational awareness. An alternative design philosophy would be to keep the operator in the loop and to view the automation as assisting the operator. The design goal is to take advantage of the benefits offered by automation, but to do so in a way that does not impede the operator's ability to perform his or her role. The achievement of this goal is difficult because it requires the combination of two disciplines, systems engineering and human factors engineering, neither of which alone can provide the solution.

This paper describes one method of function allocation. The method is a modest extension of the York method [Dearden et al. 2000], that in addition provides the means to allocate function to multiple roles, as would be found in a collaborative system. The paper's structure is the following. Section 2 gives a brief description of the method and then describes in more detail what is required as input to the method, and what is produced as a result. Section 3 provides a more detailed description of the method in particular the decision steps that are taken. Section 4 concludes by briefly describing further extensions for which there is no space to elaborate, and discusses further directions.

2 Inputs and outputs to the method

2.1 Introduction

Our method refines a set of function definitions in two steps. A function describes a unit of work that is to be performed by the proposed system. In the first step the functions are matched to a set of predefined roles. These roles capture the high level function of the agents. The aim is to decide how closely these functions fit the roles in a set of given scenarios. Functions that are entirely subsumed within a role are proposed to be "totally manual within the role" – to automate would in effect remove part of the agent's role. Functions that can be separated entirely from any of the roles, and for which it is possible to automate, are designated as "to be automated". In practice few functions fit into either of these categories. In the second step, the functions that are left are considered in more detail in order to decide what aspects of them should be automated. In the following sub-sections we describe function, role and scenario in more detail. Other inputs to the method, for example: description of the technology baseline, mandatory constraints and evaluation criteria are also required as inputs. Function, role and scenario are however the minimum needed to get a reasonable understanding of the method. The output of the method is a set of functions. These functions will either be described as manual or totally automated, or in the case of the partially automated functions, a further description is required of how the function is to be automated.

2.2 Role

Whereas in the early applications of allocation of function, the problem was to decide what people are good at and what machines are good at, recent methods have questioned this assumption. Usually the appropriateness of automation is contextually determined. It makes more sense to be specific about how compatible the tasks are with the various roles specified for the various personnel than to ask the more simplistic question: "would it be sensible for a human to do this?". In practice role is difficult to define. It makes sense to consider it as an activity that can be performed either by human or machine. Normally it will not be necessary to produce a statement for a machine's role, but there may be exceptional circumstances where it may be appropriate. For example, the envelope protection provided in a fly-by-wire aircraft could be clarified through an explicit role statement. Doing so would help to ensure that functions are not allocated in a way that prevents the machine from being able to keep the plane within the safety envelope.

An example of a role statement for the captain of a civil transport aircraft is:
> *The captain is the final authority for the operation of the airplane.*
> *He must prioritise tasks according to the following hierarchy:*
> *safety, passenger comfort and flight efficiency.*

In the context of a single seat military aircraft, an example role statement for the pilot might be:
> *The pilot is the final authority for the use of offensive weapons and*
> *for evasive manoeuvres. The pilot is responsible for mission*
> *planning, tactical and strategic decision making and co-*
> *ordination with other aircraft. The pilot is responsible for the*
> *management of systems to maximise the probability of successful*
> *mission completion.*

2.3 Functions

Work systems perform functions or *units of work*. Examples of what we mean by function include "finding the current position of the vehicle", "withdrawing cash" or "detecting a fire". In order to perform function allocation, a set of such functions is required. Although in practice, identifying and determining a function is largely a matter of expert judgement, some characteristics of a function can be determined that aid this process. A function is an activity that the man-machine system is required to be capable of performing in order to achieve some result in the domain under consideration. Functions do not contain any indication of who or what performs them [Cook & Corbridge 1997]. For example, 'key in way-point' is not a function because it implies that the operator enters the way-point manually, whereas 'set way-point' does not. Functions can be seen as being related to functional requirements as used in systems engineering. As function allocation is a design technique, a set of functional requirements will be available from which to derive functions. Care has to be taken when doing so because functional requirements often contain indications about who or what should perform the function. Such information needs to be removed.

It is important that functions are defined from a holistic and evenly balanced perspective across the systems components. Otherwise bias will adversely affect decision-making about how functions should be automated. It is therefore necessary to be aware of what viewpoint the functional requirements are taking so that this neutrality can be achieved. Functions can be defined at a variety of levels of granularity, and are often arranged hierarchically. Taking a top-down hierarchical approach can be a useful aid to function elicitation. Descending the hierarchy reveals both a decrease in the complexity of the function and in the size of the sub-system required to perform it. At the top of the hierarchy will be those functions that can only be performed by the system as a whole. Further down will be those functions that are performed by a team or department. Even further down are those functions that can be performed by a single operator with the aid of automation. Finally at the bottom are those functions that can be performed entirely by a single operator or machine. The overall pattern will depend upon the size of the system being modelled.

2.4 Scenarios

Traditionally, function allocation methods presume that the suitability of a function for automation is based on an individual function. We have noted already that this ignores the complex interactions and dependencies between activities of work. To overcome this, the method considers the allocation of functions together as groups within the context of a scenario. This allows the designer to appreciate the interactions between the functions. Scenarios should be selected that focus upon situations relevant to critical decision criteria, for example workload or situation awareness. They should represent all the functions that have been selected in a range of contexts. For example, if the system is in the field of civil aviation and workload has been identified as a decision criterion then scenarios focusing upon take-off and landing would be chosen (among others such as emergency conditions) because these periods are recognised as having a high workload.

As the method only considers those functions used within the current scenario, the designer must ensure that every function occurs in at least a minimum of one scenario and ideally several. It is important to ensure that a wide variety of scenarios are used so that confidence can be placed in the results. In order to achieve this, scenarios should be selected that cover all the normal operating conditions of the system.

There is a wide range of sources for possible candidate scenarios:

- The experience of practitioners in previous systems
- Incident and accident reports for previous systems
- Scenarios developed during the business modelling stage of the system life cycle
- Use cases and scenarios in the previous systems development documentation or training manuals

When a scenario is adopted, the description of what happens is likely to be at an event level. For example, if the scenario is describing some current possibility with the baseline architecture, these events will be expressed in implementation dependent terms. It is therefore necessary to transform these scenarios into a form in which the account of what happens is translated into functional terms. The functions being unbiased towards the implementation, will be less likely to be tied to the assumptions made in the baseline. It must be noted however that a general problem with this sort of approach is that there will be some bias.

It is unlikely that all of the candidate scenarios will be useful for the purposes of function allocation, or that they will be in the correct format. The designer therefore will have to select those appropriate and document them. Checking that they are appropriate will involve transforming each of them into a format based on functions rather than events. The scenario representation is based on a modified version of the scenario template for THEA [Pocock et al. 2001]. The template used for the scenario descriptions is shown in Figure 1.

UC#	The name of the use-case	
Scenario 1 - Each use-case can have a number of scenarios		
Environ-ment	A description of the environment within which the system is operating when the scenario occurs.	
Situation	A description of the state of the system at the start of the scenario. Are all the operators on duty, are there any known or unknown faults in the system etc.	
Sequence of events	Step	Event
	1	The main sequence of events that take place during the scenario. This includes events that happen in the environment, events that effect the system and the actions that the system must perform.
	...	
Event extensions	Step	Event
	1	Variations upon the main sequence of events are recorded as event extensions.
	...	
Scenario 2		

Figure 1 Scenario input template

The scenario as illustrated in Figure 1 is taken as a starting point. This description is then transformed. The events and event extensions headings are changed to functions and function extensions. The account of what happens in the scenario is re-expressed using the set of functions rather than the event descriptions. This process can be a useful check that the functions are expressed at an appropriate level and whether they incorporate too much implementation bias. The functional scenarios are then used as the basis for the two decision procedures.

2.5 The nature of the output

Traditionally function allocation methods provide two allocation options H (human) or M (machine). It is assumed therefore that functions are at a very low level so that such binary decisions can be made. In many cases the designer's interest lies with those higher level functions that require some form of collaboration between operator and machine. It is the determination of where this automation boundary lies that is of critical importance. Rather than working with numerous low-level functions to determine this boundary, it is easier to work with the parent functions and to declare them as partially automated. Of course there are many ways in which the operator and machine can interact in the fulfilment of a function. There is therefore a need for the designer to specify how the partnership will operate.

One possible approach is to use a classification that defines levels of automation. [Sheridan & Verplanck 1978] suggest a classification of levels in which decisions and control pass progressively from the human to the machine. Later authors have produced alternatives. Kaber and Endsley provide ten levels of automation [Kaber & Endsley 1997]. Billings suggests seven levels of management automation [Billings 1991]. For each function that the designer defined as being partially automated, there would be an indication of what the level of automation would be. The difficulty with this approach is that defining which level of automation is required does not necessarily define what it means for the particular function to be implemented. Another problem with such an approach is that some solutions for functions may not fit into any of the classifications.

The approach we take is to provide a conceptual framework for thinking about partial automation. The framework makes use of the [Malinkowski et al. 1992] framework for adaptive systems and is capable of expressing all types of automation provided by the various classifications. A function is first split into four components: Information, Decision, Action and Supervision. Each component is further split into a number of elements, each describing a particular aspect of the function. The designer specifies which role is responsible for performing that aspect of the function. If the element is not applicable within the context of the function then it is marked not applicable (n/a).

The elements in the Information component cover issues such as which role integrates the information required to carry out the function, and which role is responsible for initiating a response. The Decision component covers such issues as which role proposes what plan/action to take, evaluates it, modifies it and selects one, if there is more than one possibility. The Action component covers which role carries out the action. The Supervision component covers such issues as which role monitors the performance of the action, identifies exceptions, and revokes the action if necessary.

This framework (called IDA-S, see [Dearden 2001]) allows the designer to express precisely how the function is implemented in terms of the various roles that are responsible aspects of the function. This is done by placing a role identifier in an appropriate cell in the IDA-S template. In the case of the function: "calculate point to point information", one possible solution might be that the navigator selects two points upon the chart, that is the current fix and next way-point. The navigation system calculates the distance, time and bearing to the way-point. In Figure 2, the required roles are placed in the appropriate cell of the grid. In some cases, where more than one role appears to be appropriate, they may all be placed in that cell. A complication, such as this, may suggest that the function has not been sufficiently refined at the stage of selection, and may suggest a need to revisit the function

Allocation	Information		Decision		Action	
Planning the response	Collect		Propose		Approve	
	Integrate		Evaluate			
	Configure		Modify			
	Initiate response		Select			
Supervise ongoing execution	Monitor progress		Identify exceptions		Revoke authority	
Supervise termination	Determine output content		Identify completion		Stop process	
Action	Execute actions					

Figure 2: The IDA-S template

definitions and the scenarios. The IDA-S definition clarifies how to develop an implementation that satisfies the requirements.

3   The method: two steps and consolidation

3.1 Totally automated or wholly manual?

Once roles, functions, scenarios, and so on, have been defined the first step involves deciding which functions can be totally allocated to one of the roles (these roles may be system or human). Each scenario is considered in turn and the functions, that are employed within the scenario, are identified. The designer bases decisions about automation of these functions on their use in the context of the scenario under consideration. Suitability for total automation is not based solely upon the technical feasibility of a solution. It is also based upon the function's relation to the roles. If a function is not seen to be separable from an operator's role then it cannot be totally automated. Doing so would interfere with the operator's ability to do the job effectively. There are two dimensions to the trade-off.

Firstly, each function is considered in relation to the feasibility of automating it. The concern here is with the cost and technical possibility. Secondly, the function is matched against the set of roles. The roles are likely to continue to be refined as the design evolves. In Figure 3, two roles R1 and R3 are relevant to the given scenario. In practice, a function may be separable from all roles, and technically feasible and cost effective to automate, in which case the function may be totally automated. Alternatively it is possible that the function maps entirely to one of the roles, and is infeasible to automate, in which case the function is totally performed within that role. In most cases however functions fit into neither category. In this situation the function is to be "partially automated". There are usually a number of ways in which partial automation may be achieved, and this is the subject of the second trade-off. The first decision procedure is made traceable through the matrix in Figure 3. Notice that a function may appear in more than one row because it relates to several roles, but may only appear in one column because feasibility to automate is invariant.

State of automation research vs. relation to role	Role	Existing with immediate access	Existing in competitor systems	Low risk/low cost R&D	High risk or high cost R&D	Infeasible
Separable	ALL		Sol1			
Role related information or control	R1 R3					
Role critical information or control	R1 R3					
Central to role	R1 R3					

Figure 3. The first trade-off

When this process has been completed, a set of functions has been produced that can either be totally automated or are entirely subsumed within a role. These functions go no further in the process. The remaining functions go forward to the next stage of the allocation of function process. These functions are represented using the IDA-S framework defined above.

3.2 Candidates for the second step

The second decision procedure is concerned with comparing alternatives with a "baseline" solution in the context of a scenario. Often the baseline is the current design, but it may be that, in the case of a completely new concept, there may be a more conservative design possibility. It is assumed here that an initial process is engaged in where the designer constructs a number of alternatives for each function listed from the scenario. These alternatives may allocate different roles to elements of the IDA-S template. For example, in the case of "calculate point to point information", we could consider one solution in which the navigation system might propose alternative routes from which the navigator selects the most appropriate, while in another no such choice is provided. The reason for choosing particular representations may be random or based on some assumptions about the abilities of the off-the-shelf technologies available to the project. The options for all these functions are then rated in relation to a criterion such as workload or performance in comparison with the "baseline" design.

When all the options for all the functions have been produced, and the primary concern identified, the options are entered into a second matrix (see Figure 4). The design options used in the baseline should also be included in the cell that is labelled "no significant improvement in primary concern". The alternative solutions are then placed in the table. Two criteria decide where the solution should fit. The first depends again on the feasibility of the particular solution, how easy will it be to implement with existing technology. The second requires a judgement about the effect of the solution, in the context of the scenario, in terms of the criteria (workload, performance, situation awareness). The judgement here is whether the function solution causes an improvement or deterioration to the primary criterion and the consequent effect on the other criteria. All these judgements are carried out in relation to the baseline design. It would be expected that some solutions will do better while others will do worse. It is intended that the designer uses expert judgement, but it could be that the situation requires a more careful human factors analysis of these decisions.

Each potential solution is placed in the matrix. At the end of the process it will be possible to derive a set of best candidates that are the solutions that are most favourable in relation to the criteria and are technically feasible. This process is achieved by searching from the top left of the matrix selecting new design options. If a design option for a function is selected, then all other options for that function are deleted from the table. If a design option is selected from the 'high risk research and development' column, then an alternative, low risk solution should also be considered as a 'fall-back' position. Hence in Figure 4 two options are provided for a function F.1.2.4. Both are implementable because they are available on competitor systems. However the second solution is preferable because while it results in an improvement in performance compared with the baseline solution, it has no deleterious effect on any of the secondary concerns such as workload.

After a number of options for functions have been selected, the designers should re-evaluate the scenario and consider whether or not the primary concern should be changed as a result of the decisions made so far. For instance, consider a scenario in which high workload is the primary concern. If new partially automated solutions are selected, that significantly reduce the expected workload, then a different concern such as performance or situation awareness may now be more significant. If the primary concern is changed, then options for the remaining functions are re-arranged in a new matrix reflecting the changed priorities. Option selection then proceeds as before, starting with the options that provide the greatest improvement for the new primary concern. This procedure iterates until one design option has been selected for every function.

One possible outcome of the procedure is that some functions cannot be successfully allocated, without making use of options from the high risk research and development column, or from a row involving a 'large deterioration' with respect to a secondary concern within the scenario. If this occurs frequently, and cannot be solved by generating alternative design options, this may indicate a need to review the system requirements, or to review assumptions about the number and role of human operators in the system.

When a design option for each function has been selected, the scenario is re-analysed using the proposed allocations as the set of baseline designs. The purpose of this re-analysis is to identify any new functions that may be an emergent consequence of the new design. Such functions could include, for example, a requirement to co-ordinate two separate functions that control the same system resources (e.g. 'terrain following' and 'missile evasion' both use the aircraft's control surfaces). Design proposals for the partial automation of these functions are made.

Primary concern:	Performance			
State of automation vs level of improvement	Suggestion is immediately available	Available on competitor systems	Low risk, low cost R&D	High risk or high cost R&D
Large improvement in primary concern, no deterioration in secondary concerns				
Improvement in primary concern, no deterioration in secondary concerns		(6) F1.2.4 Sol 2		
Improvement in primary concern minor deterioration in secondary concerns		(10) F1.2.4 Sol 1		
Improvement in primary concern, large deterioration in secondary concerns				
No significant improvement in primary concern				
Large deterioration in primary concern				

Figure 4: Identify partially automated functions

If new functions are identified, then designers must consider whether their impact upon performance, workload and situation awareness is acceptable. If the emergent functions do create an unacceptable situation, then the selection matrix is revisited. This may result in changing the level of automation, or may result in changed selections for the original functions. Hence, if emergent functions are identified then steps of the process dealing with these are repeated, i.e. feasible design options are suggested for partially automating the emergent functions and the optimum choice is selected.

3.3 Consolidation

Once the functions within each scenario have been allocated any contradictions of allocation across scenarios are resolved. This is done, either by changing one of the allocation decisions so as to resolve the conflict, or by allowing the function allocation to be dynamic. Allocation conflicts may be resolved by choosing solutions that are most favourable *across the scenarios* using the primary and secondary criteria of the second trade-off. There may be situations where this global compromise is difficult, if not impossible, to make and this may occur in circumstances where the level of automation does not have to remain fixed. Components of IDA-S allocations can be transferred from one role to another during the activity supported by the system. Dynamic function allocation refers to the process of redistributing functions amongst roles, with the goal of improving overall system performance. This redistribution typically occurs in response to a change in the environment or a change in the state of one of the agents assuming the role. The designer must decide how the allocation of function changes, and the extent to which the operator is involved in this process. In practice, the change-over can be seen as another function to be refined, in the same way as any other function that emerges during the process.

When a function is redistributed amongst human agents it is often referred to as workload sharing. When a function is redistributed between human and machine it is often referred to as adaptive automation. Redistribution can either be human-triggered or machine-triggered. Much current research focuses on dynamic allocation in response to changes in workload. Triggers for reallocation also include, for example:

- Critical system events – adaptive automation is triggered if a system error occurs
- Performance measurement - degradations in human vigilance invoke adaptive automation
- Psychophysical assessment - dynamic assessments of human physiology drive the decision as to whether to automate
- Behaviour modelling - control allocations are used to achieve a predetermined pattern of overall system functioning [Parasuraman & Mouloua 1996].

A detailed description of this stage of the process can be found in [Johnson et al. 2001a].

4 Extensions and conclusions

There has not been space to discuss the work that has been done to integrate the method with notations and representations that are relevant and comprehensible to systems engineers. The main steps here have been to develop an extension to the use-case mechanism of UML [Rumbaugh et al. 1999] in order to provide a more readily accessible representation of scenarios, and to use activity diagrams from UML to describe IDA-S patterns. The difference between the templates described in this paper and activity diagrams is that a further commitment needs to be made about the sequence in which information, decision, action and supervision takes place which is not contained in the IDA-S template. The UML proposal therefore gives further guidance to the systems engineer about how to implement the functions. More information about this is contained in the fuller paper [Johnson et al. 2001a].

An important issue has been to keep the method sufficiently procedural and straightforward that it can be used in practice. For this reason we are reviewing the complexity of the IDA-S template. We are looking at alternatives that provide attributes that are simpler to understand. An area where the method is perhaps too simplistic and untried is the area of adaptive automation. It may be possible to do better than assume (1) that automation adapts by function substitution and (2) that in all cases the ideal result is achieved. In practice both assumptions may be false. As noted by [Sperandio 1978], in discussions of air traffic control, different strategies are adopted depending on the number of aircraft in the air space. Sequences of functions making up procedures rather than individual functions are substituted. It may also be appropriate that dynamic mechanisms should be prepared to shed certain less critical functions in the face of hard deadlines. Both these issues are discussed in a little more detail in [Harrison & Bernat 2001].

Acknowledgements: We would like to acknowledge the contributions of Andrew Dearden, Colin Corbridge and Kate Cook to this work and thank Shamus Smith for reading draft versions. It has been funded successively by BAE SYSTEMS and DERA CHS.

5 References

[Bainbridge 1987] Bainbridge, L. Ironies of automation. In Rasmussen, J, Duncan, J. & Leplat, J. (eds). New Technology and Human Error, pp 276-283, John Wiley & Sons, 1987.

[Billings 1991] Billings, C.E. Human-Centered Aircraft Automation. Technical report number 103885. NASA AMES Research Center, 1991.

[Cook & Corbridge 1997] Cook, C.A. & Corbridge, C. Tasks or functions: what are we allocating? In E. Fallon, L. Bannon & J. McCarthy (eds.) ALLFN'97 Revisiting the Allocation of Function Issue: New Perspectives, pp. 115-124. Louisville KY: IEA Press, 1997.

[Dearden et al. 2000] Dearden, A., Harrison, M.D. & Wright, P.C. Allocation of function: scenarios, context and the economics of effort, International Journal of Human-Computer Studies 52(2) 289-318, 2000.

[Dearden 2001] Dearden, A.M. IDA-S: a conceptual framework for partial automation. To appear in: Blandford, A., Vanderdonckt, J. and Gray, P. (eds.) People & Computers XV – Interaction without Frontiers. Proceedings of IHM-HCI 2001. Springer. Berlin, 2001.

[Fuld 2000] Fuld, R.B. The fiction of function allocated, revisited. International Journal of Human-Computer Studies. 52(2) 217-233, 2000.

[Grote et al. 2000] Grote, G., Ryser, C., Wafler, T., Windischer, A. & Weik, S. KOMPASS: a method for complementary function allocation in automated work systems. International Journal of Human Computer-Studies. 52(2) 267-288, 2000.

[Harrison & Bernat 2001] Harrison, M.D. & Bernat, G. What can dynamic function allocation learn from flexible scheduling. University of York. DIRC working paper. 2001. www.cs.york.ac.uk/~mdh/papers/harrisonb01.pdf.

[Johnson et al. 2001a] Johnson, P.D., Harrison, M.D. & Wright, P.C. An enhanced function method. University of York www.cs.york.ac.uk/~mdh/papers/johnsonhw01.pdf.

[Johnson et al. 2001b] Johnson, P.D., Harrison, M.D. & Wright, P.C. An evaluation of two methods of function allocation. People in Control IEE Press. Conference Publication No. 481. 178-183. 2001.

[Kaber & Endsley 1997] Kaber, D.B. & Endsley, M.R. The combined effect of level of automation and adaptive automation on human performance with complex, dynamic control systems. In Proceedings of the Human Factors and Ergonomics Society 41st Annual Meeting. pp 205-209. Santa Monica CA.: Human Factors and Ergonomics Society, 1997.

[Lim & Long 1994] Lim, K.Y. & Long, J.B. The MUSE Method for usability engineering. Cambridge University Press. 1994.

[Malinkowski et al. 1992] Malinkowski, U., Kuhme, D.H., Schneider-Hufschmidt, M. A taxonomy of adaptive user interfaces. In Monk, A., Diaper, D. & Harrison, M.D. eds. People and Computers VII, Proceedings of HCI'92. pp 391-414. Cambridge University Press, 1992.

[Older et al. 1996] Older, M., Clegg, C & Waterson, P. Report on the revised method of function allocation and its preliminary evaluation. Institute of Work Psychology, University of Sheffield, 1996.

[Older et al. 1997] Older, M.T., Waterson, P.E. & Clegg, C.W. A critical assessment of task allocation methods and their applicability. Ergonomics, 40: 151-171, 1997.

[Parasuraman & Mouloua 1996] Parasuraman, R. & Mouloua, M. (eds). Automation and human performance: theory and applications. Lawrence Erlbaum, 1996.

[Pocock et al. 2001] Pocock, S., Harrison, M.D., Wright, P.C. & Johnson, P.D. THEA: a technique for human error assessment early in design. IFIP TC 13 International Conference on Human-Computer Interaction. IOS Press, 2001.

[Rumbaugh et al. 1999] Rumbaugh, J., Jacobson, I. & Booch, G. The unified modelling language reference manual. Addison Wesley. 1999.

[Sheridan & Verplanck 1978] Sheridan, T.B. and Verplanck, W.L. Human and computer control of undersea teleoperators. Technical report. Man-machine systems lab, Dept of Mechanical Engineering, MIT, Cambridge, MA, 1978.

[Sperandio 1978] Sperandio, J.-C. The regulation of working methods as a function of workload among air traffic controllers. Ergonomics 21:195-202, 1978.

The Processes to Manage (and Minimise) the Human Risk in Complex Systems

Brian Sherwood Jones
Process Contracting Limited
Prestwick
Scotland
Jonathan Earthy
Lloyd's Register of Shipping
London
England

Abstract

This paper explores the use of the Human-System process (HS) model in managing the human-related risk in complex systems. The HS model is a proposed ISO Publicly Available Specification (PAS) [ISO PAS tba:2002] A specification for the process assessment of human-system issues. It presents a view of the system life cycle with an emphasis on the identification and handling of issues related to people (users and other stakeholders). The model is focused on system acquisition and operation but includes processes related to Human Resources (e.g. selection and training). It is intended for use in process assessment and improvement, but could also support planning and the assessment of competence. A process assessment approach to IEC 61508 [IEC 61508:1998] has been proposed for software-related processes [Benediktsson *et al* 2001], and is recommended here as a validated means of assessing organisational capability to deliver systems in a user-centred manner. The relationship between the processes in the HS model and those required by Health and Safety is discussed. The HS model is also proposed as a means of addressing compliance with Regulations across sectors in a consistent way.

1 Introduction

This paper describes recent work on process-based approaches to assurance of system properties (including but not exclusively, safety) and situates these approaches in the context of safety assurance. The thesis is that process-based approaches have a key role (but not an exclusive one) in providing assurance of safety-related or safety-critical systems.

The focus is on the human-related risk in complex systems. A system is considered as a work system, rather than a piece of equipment. ISO 6385 [ISO 6385:1981] defines a work system as *"a combination of people and working equipment, acting together in the work process, to perform the work task, at the work space, in the work environment, under the conditions imposed by the work task"*. From this point of view the authors contend that human-related risks are the major system risks. The paper has a strong emphasis on the need to integrate assurance of human-system (HS) issues with other engineering activities.

The aim of the work discussed is to achieve 'quality in use', colloquially referred to in this paper as operability. Quality in use is defined [ISO 9126:2000] as *"the capability of a system to enable specified users to achieve specified goals with effectiveness, productivity, safety and satisfaction in specified contexts of use."*

The paper discusses the nature of safety assurance, the types of indicator that might be used, and their strengths and weaknesses. In particular, indicators are examined with respect to the extent to which they are lead indicators, i.e. they predict problems, as opposed to a lagged indicator, where we are wise after the event.

The basics of the process-based approach are then described, with an outline of its potential role in safety assurance. The scope and nature of a process model to encompass human-system (HS) issues is presented, together with the background in process-based developments in system and software engineering and their application to safety-critical systems.

Process-based approaches are then discussed in relation to other types of indicator that are used in safety assurance.

The paper draws conclusions concerning the use of a process-based approach. In particular, it concludes that a process-based approach has the potential to become the mainstream approach to safety assurance as regards human aspects of systems (and possibly for all aspects, though this goes beyond the scope of the paper).

2 The nature of safety assurance as prediction

Turner [Turner 1978] was one of the first to try and identify indicators that might predict man-made disasters. He proposed the idea of an 'incubation period' prior to a disaster, and that the characteristics of an incubation period could be identified. The identification of the characteristics of potential disaster, or likely success, has occupied many minds since. Lead (rather than lagged) indicators are vital to safety assurance. A range of diverse indicators will probably always be required. Some candidate indicators are:

- The 'three P's' of Product, Performance and Process characteristics,
- Compliance with Regulations,
- The extent to which best practice methods are being followed,
- Organisational characteristics such as safety culture,
- The vigour (and rigour) of risk management,
- The quality of design decision making.

The merits of process over the other P's of product and performance are clear. Specification and assurance of product characteristics will always lag behind new technology, cannot consider system-level effects beyond the item of equipment, and are sensitive to the context of use. Performance is inevitably a lagged indicator. Compliance with Regulations is clearly necessary, but offers no assurance of being sufficient. Furthermore, the move to open-textured Regulation makes assurance more judgmental. The other indicators listed above are widely used in day to day project working, but present difficulties in providing assurance.

3 The process-based approach and its potential for safety assurance of HS issues

The attraction of a process-based approach to safety assurance lies in the potential for lead indicators. The work (led by the software community) on defining process in a way that can be measured can enable this potential to be realised. This section describes the process-based approach and summarises the Human-System (HS) model (being developed as ISO PAS tba *A specification for the process assessment of human-system issues* (ISO PAS 'HS')). It then discusses the scope of the HS model in relation to the demands of safety assurance for human-related risks.

3.1 The process-based approach

Processes are defined at the level of what is done to develop and operate a system. They are specified through methods, techniques, work instructions, etc. A process has a purpose and fulfils a business requirement. A technical or management task that contributes to the creation of the output (work products) of a process or enhances the capability of a process is called an activity or a practice. The elements of the format used to describe processes in ISO 15504 [ISO TR 15504:1998-2] are listed in Table 1. These have been followed in ISO TR 18529 *Human-centred lifecycle process descriptions* [ISO TR 18529:2000] and in ISO PAS 'HS' [ISO PAS tba:2002].

A disciplined evaluation of an organisation's processes against a model is called process assessment. Process assessments generally focus on identifying improvement priorities (i.e. they are formative evaluations). Action taken to change an organisation's processes so that they meet the organisation's business needs and achieve its business goals more effectively is called process improvement. Process assessment seeks firstly to establish whether processes are performed successfully and secondly the degree to which processes are under control. A capability scale (an ordinal scale of types of control) is used in this assessment. Table 2 describes the six-level ISO 15504 capability scale.

Name	Description of component
Process number	For precise reference
Process name	Summary of the process
Purpose of process	What is done by the process
Outcome	Why it is done, the result of successful application of a process
Practices	What is done to fulfil the purpose
Practice number	For precise reference
Practice name	Summary of the practice
Description of practice	What task is performed
Work products	The items used and produced by the process including the following: pieces of information, documents, hardware, software, training courses, awareness in individuals.

Table 1: the components of a process have precise definitions in ISO 15504 to enable rigorous assessment

Level	Description
0	No achievement of results from a processes or processes
1	Performance in an ad hoc manner
2	Monitoring of time and product quality
3	Use of defined corporate procedures and infrastructure
4	Use of statistical control
5	Optimisation of each process to meet current and future business needs

Table 2: Capability levels in ISO TR 15504 *Software Process Assessment* are used to describe how well processes are performed

A process assessment examines the evidence for the performance of practices and the existence and quality of work products. Process attributes are features of a process that can be evaluated, providing a measure of capability in doing the process. However, the final decision as to the degree of performance of a project is based on the degree to which the outcomes are achieved.

3.2 The HS Model

The HS model [ISO PAS tba:2002] comprises 3+1 processes that address issues associated with people through the system life cycle. Process HS.4 is considered to be less well-substantiated than HS.1 to HS.3 and so appears as an annex. The expectation in the (very wide) community that developed the standard is that assessments are likely to focus on the level 0-1 capability distinction. The processes are described in Table 3. The relationship between the processes and their sub-processes is outlined in Figure 1.

HS.1 Life cycle involvement.
This process anticipates the particular HS issues at specific stages of the life cycle. It makes the system life cycle efficient by addressing people in the stage enabling systems for the system. Its is to consider the interests and needs of the individuals and/or groups that will work with the system. It is achieved through performance of five sub-processes which are in general grouped according to the example stages provided in Annex B of ISO CD 15288 [ISO CD 15288]. However, in order to create meaningful groups of HS activities the utilisation stage is split between the early stages (installation and transition to use) and the mainstream use of the system (operation and support of the system).
HS.2 Integrate human factors.
This process ensures that HS issues are addressed by the appropriate stakeholders. It reduces life cycle costs by ensuring that design for people is used within the organisation. The purpose of the Integrate human factors process is the satisfactory deployment of human-system processes for a system. It is achieved through performance of eight sub-processes.
HS.3 Human-centred design.
This process enables user centred technical activity to be focused appropriately. It contributes to a better system by designing for people who use the system of interest in its context of use. The purpose of the Human-centred design process is to apply HS processes and HF data as appropriate in order to ensure the usability of the system throughout its life cycle. It is achieved through performance of four sub-processes.
HS.4 Human resources.
This process provides the means to resolve issues by means of the human part of the system, rather than the equipment-centred part. It ensures the continued delivery of the correct number of competent people required to use the most suitable equipment. The purpose of the Human resources process is for usability to be achieved in the most timely and cost-effective manner by provision of the correct number of competent users. It is achieved through performance of four sub-processes.

Table 3: The HS model has four processes. The first addresses the needs of the life cycle. The second addresses the various interfaces to the organisation. The third comprises the iterative technical cycle, and the fourth addresses the timely delivery of trained operators.

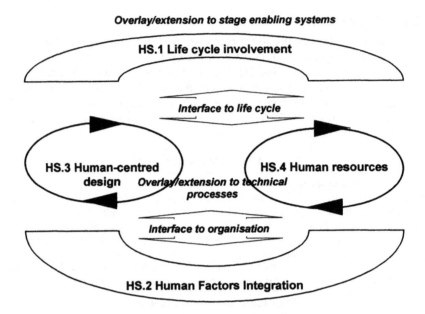

Figure 1 Process HS.1 manages the interface between the technical processes and the life cycle, while HS.2 manages the interface to stakeholders.

3.3 The scope of assurance required for HS issues

The scope of work necessary to prevent human error is characterised by its breadth, embracing the safety management systems of both supply and operation organisations, and coverage of the whole work system. For example, the HSE framework [HS(G) 65: 1991], [HSG48: 1999] for setting (and presumably assessing) performance standards covers organisational factors, job factors and personal factors (where job factors include plant and substances, procedures and the premises). This breadth of coverage, where (for example) the whole process of equipment design is a sub-sub-set, brings problems of complexity and scale to any attempt to describe the processes involved. Indeed, minimising complexity and scale was perhaps the major issue confronting the development of the HS model. In that work, the area that was found to be the least well-developed was the human resources area; there was considerable background work to describe "putting the machine in front of the user" but much less of relevance to describe "putting the user in front of the machine".

The nature of activity to prevent human error is frequently seen as the province of ergonomics specialists. It is currently the case that the enumeration of human error potential requires deep skills possessed by a small group of specialists. However, the scope of human error mitigation activity demands that it cannot be a specialist activity. One of the characteristics of recent process-based work such as

ISO PAS 'HS' is that it has removed specialist "push" from the description of the scene, and has concentrated on the outcomes to be achieved. The development of a process model that can be used for rigorous assessment has also demonstrated that the "soft, fuzzy" attribution given to user-centred design is no longer valid (or rather is no more valid than for any other branch of engineering).

4 Process in the context of safety assurance

The sections below situate process indicators in the context of other (possibly more established) indicators used for safety assurance. The intention is to identify the role that could be played by each.

4.1 Process and methodology

The 1980's saw the development of many heavyweight methodologies for software development, human factors, and for system engineering. These have since been found to pose difficulties in practical implementation. They require tailoring, but provide no guidance as to what constitutes satisfactory tailoring as opposed to non-compliance, and still posed significant cost penalties.

Since then there have been two developments. The first, the development of lightweight methodologies, such as eXtreme Programming, poses the risk of "hack and hope", and is not appropriate to safety-critical systems. The other development has been the rise of 'process' rather than method. The aspiration of the process-based approach has been to offer flexibility in application and brevity in description whilst still providing rigour in the assessment of outcome or activity.

The arrival (or imminent arrival) of process-based standards offering rigorous assessment for system engineering, software engineering and operability offers the safety community the potential for a more integrated approach. Some promising investigations in this area have already been undertaken. [Benediktsson et al 2001] have proposed a correspondence between the safety integrity levels (SILs) of IEC 61508 and the capability levels (CLs) of ISO 15504, (both built on ISO/IEC 12207 *Software lifecycle processes*). They considered the ISO 15504 reference model to be an appropriate framework for assessing safety-critical software processes, and provided empirical support for this based on the SPICE trials. The reference model of ISO 15504 provides a way to define and assess *what* is to be achieved. IEC 61508 describes the what but also the *how*. Benediktsson *et al* showed that the greater demands of higher SILs can be expressed as higher levels of process capability. An assessment model at the 'how' level of safety aspects of systems would be both possible and necessary and would enable safety to be considered in line with other aspects of systems and software. The benefits to be obtained are considerable and are mostly concerned with reducing cost. The use of a process-based approach would:

- Allow safety to take advantage of Off The Shelf (OTS) assessment methods, tools etc.

- Permit integration with other assessments, including assessment of human-system issues, system engineering processes and software engineering processes.

- Provide an OTS framework for Process Improvement.

- (Arguably) provide a more systematic and rigorous approach to assessment.

There has been long-standing discussion as to whether people are part of "the system". The ability to draw on ISO 15288 and ISO PAS 'HS' would provide assurance of the full work system - particularly where software cost avoidance has moved safety criticality from equipment to people.

In brief, the potential for the process-based approach to underpin both project execution and safety assurance (of both technical and human-system issues) offers the benefits of simplicity and lower cost.

4.2 Process and life cycle

Processes should not be confused with the stages of a life cycle. Processes are enacted at more than one stage in the life cycle, and it may be useful to think of them as essentially continuous through the life cycle. In particular, the life cycle processes in ISO PAS 'HS' have a role throughout the system life cycle. At each life cycle stage, it is necessary to look ahead to future stages ("how will we deal with disposal?"), and to check that the requirements and constraints generated by previous stages have been met ("is the original concept still valid?"). This is shown in Figure 2 below.

Stakeholder	HS1 Life cycle involvement process; Human-System Issues in..				
	HS1.1 Concept	HS1.2 Development	HS1.3 Production and Utilization	HS1.4 Utilization and Support	HS1.5 Retirement
Conceiver	**Needs, Concepts, Feasibility**	Consistency, Viability	Consistency, Viability	Consistency, Viability	Consistency, Viability
Developer	Compatibility Feasibility	**Engineering, Solutions, Practicability**	Consistency, Viability	Consistency, Viability	Consistency, Viability
Deliverer (HR, trainer)	Compatibility Feasibility	Compatibility Feasibility	**Manufacture, Roll-out, Installation**	Consistency, Viability	Consistency, Viability
User	Compatibility Feasibility	Compatibility Feasibility	Compatibility Feasibility	**Operation, Support, Validation**	Consistency, Viability
Disposer	Compatibility Feasibility	Compatibility Feasibility	Compatibility Feasibility	Compatibility Feasibility	**Reuse Archiving Destruction**

Figure 2 The continuing nature of processes and stakeholder involvement

Perhaps the stakeholder that needs explanation in Figure 2 is the Human Resources (HR) personnel provider and trainer. Most of the other stakeholders are likely to be most concerned with getting equipment off the drawing board and into operation. However, the Human Resources process (HS.4) to deliver Suitably Qualified and Experienced Personnel (SQEP) operators and maintainers in a timely manner must fully engage with the equipment processes.

The emphasis between the outcomes of a process will vary depending on the stage at which it is performed. This variation in emphasis will in turn affect the conduct of the practices that comprise the process. The effect of stage and project context on the performance of processes and practices is one of the main differences between process models and methods/methodologies for system development.

The process models developed have the ability to provide assurance of life cycle issues - a potential defence against latent errors [Reason 1990].

4.3 Process, culture and maturity

This section examines the relationship between organisational culture and maturity, and discusses the role of process in relation to them. It then goes on to discuss the possible role of user-centred design as a healthy culture.

To the person starting to work in an organisation, it is clear whether the organisation has "got its act together", or whether there is fragmentation, poor information flow, vested interests and re-work. Experience - within or across organisations - can frequently lead to adaptation to the second situation. There have been a number of attempts to formalise intuitions about culture, particularly because of its safety connotations. Turner [Turner 1978] proposed that a cultural disruption occurs during the incubation period whereby groups and individuals gradually come to develop and rely on a mistaken view of the world. He associated this with the development and maintenance of 'bounded decision zones' which lead to the risk that contingencies are missed, ignored or under-estimated. Vaughan [Vaughan 1996] conveys the ease of being ensnared by the production of culture, and gives a compelling account of the subtlety of its interaction with the culture of production and structural secrecy.

Westrum [Westrum 1997] proposed types of organisational climate, ranging from the pathological, through the bureaucratic to the generative. The distinguishing features are the quality of information flow and the vigour of questioning and enquiry. Perhaps the key indicator is the treatment of bad news, which ranges from 'messengers are shot' through 'messengers are listened to if they arrive' to 'messengers are trained'. The first author's personal experience is that it is easy for people in an organisation to identify where they are on the scale with some consistency. However, it would be a difficult judgement call to specify a required point on the pathological/generative scale for a particular SIL.

Figure 3 offers a caricature of a questionnaire to investigate safety culture in a formal context. The difficulty of obtaining realistic attitudinal information in a context with serious consequences should be apparent. There are a number of excellent schemes to encourage safety culture in a formal manner. These are used by organisations that are already safe to improve matters further, but the target organisations may alternatively be able to buy OTS paper trail generators that meet the letter without the spirit.

Attitudinal indicator	Tick if you agree	Consequences
Safety is my highest concern	☐	Win contract, keep working
Safety is for wimps	☐	Dismissal, Stop Work Order, lost contract

Figure 3 - Attitudinal indicators in formal safety assurance. The official form may not have the consequences itemised, but word gets around..............

The process-based approach started with the concept of organisational plateaux of performance [Paulk *et al* 1993], termed maturity levels. These represent similar behavioural syndromes to Westrum's climates. In essence, a *gradus ad Parnassum* was proposed. This starts with *ad hoc* activity where results are achieved by individuals, and moves up through levels of greater process definition, control, management and optimisation. There is an assumption that formalisation is "a good thing". In terms of starting software process improvement, the idea that "we need to get to the next level" has a compelling simplicity. There are many implicit assumptions about organisational characteristics and evolution, with something of a "one size fits all" set of Key Process Areas to move from one level to the next.

The Human-centredness scale - a supplement to the process definitions in the Usability Maturity Model precursor to ISO TR 18529 [Earthy 1998] presented a similar set of steps, but with greater account of cultural issues. A key step of relevance to the safety community is from Level C (Implemented) where human-centred processes are carried out and produce good results but where "too late to change that" is a frequent problem, to Level D (Integrated) where the life cycles are managed to ensure that the results of human-centred processes are visible in all relevant work products.

The practical disadvantages of organisational maturity (as opposed to process capability) relate to the "one size fits all" character and the assumptions about evolution. Knowing that your organisation is at a specific level does not help to diagnose any root causes, or to identify priorities in terms of process improvement for your specific business at that time. It is probably better - though perhaps less attractive than taking an organisational maturity level as a target - to bite the bullet and identify the processes that are vital to the business, set target capability levels and base process improvement on the gaps found.

Although there is a clear organisational element in the achievement of safety culture, we would argue that an appropriate 'wiring diagram', though necessary, is not sufficient, and that process measures are more powerful than structural measures in finding out whether or not management processes 'work'.

The thesis offered is that indicators of safety culture are vital to safety assurance but have to be used informally for the purposes of self-assessment and improvement rather than vendor assessment, regulation, or formal management. Further, they face the inevitable difficulty of the self-deception they are there to resolve. Maturity models developed for software development or for human-centredness have similarities to cultural models, but no particular attraction to safety assurance. Capability evaluation has a role to play that complements safety culture, being suitable for formal purposes and offering a role in self-assessment and process improvement that would support cultural improvement or maintenance.

Simply put, safety culture metrics are best used informally for improvement programmes, while capability evaluation is appropriate for formal assessments.

An entirely separate point about the relation between user-centred design and culture can be made - that user-centred design (which can be measured for formal purposes) promotes a safe culture (which can't). The user community is the place to spot - and stop - resident pathogens [Reason 1990]. Westrum's words (op. cit.) make the point eloquently:

"...Similarly, the safe system is one whose design is carefully thought through and tested, typically in conjunction with users. Its design takes into account the activities and habits of social groups that are going to use the system. It brings these groups into the design process, either through field studies or by placing them on design committees. Even the best design requires thorough training for the users and an open line of communication between the users and the designers.

Once the system is operational in the field, it needs to be monitored by the design group. Some kinds of hardware and software problems will become apparent only through use in the field. The problems that arise need to be carefully studied and rapidly corrected. But above all, somebody needs to be paying attention."

4.4 Process and safety principles

To some extent, this section is a generic version of the usual project discussion on the mapping between Product Breakdown Structure and Work Breakdown Structure - the relationship between "doing the right things" and "achieving the correct design intent". This distinction is of particular concern to operability and safety management, where it can be the case that the right things are being done by specialists, but mainstream design engineers are ignoring them. Operability (or safety) cannot be achieved in a context-free fashion; the specific targets and meanings for generic statements of design intent must be determined (as well as met) by the work programme.

Safety principles, design principles, safety criteria are phrases that are widely used in standards and guidance. The attraction of principles is their brevity. This is also their weakness - their abstraction and loss of context makes interpretation difficult and trade-offs between conflicting principles impossible. Design principles such as conservative design, segregation, diversity, redundancy inform the designer, but the assessment of achievement is highly context-specific. Functional safety goals (such as appear at the top of fault trees) inform the design, but provide very limited guidance. Principles can often end up including both work programme requirements e.g. "a task analysis should be carried out, procedures should be produced and validated" and design intent e.g. "user interface design should follow good ergonomics practice". Three examples of principles related to operability are given in Table 4.

Safety of machinery - Ergonomic design principles. Part 1. Terminology and general principles [BS EN 614-1:1995]

This standard called up by the Safety of Machinery Regulations sets out a high level goal of achieving an efficient, healthy and safe interaction of operators with work equipment. It observes that a task analysis is required to meet the goal of designing the work system to be consistent with human capabilities, limitations and needs - otherwise there is no explicit linkage between the design principles given and the specification of ergonomics tasks to be performed. The design principles specify human-machine matches to be achieved taking task characteristics into account and are at two levels of abstraction. For example:
- "Work equipment shall be designed with proper regard to the body dimensions of the expected population of operators..." and
- "The type, location and adjustability of any seating provided shall be appropriate to the dimensions of the operator, and to the tasks the operator performs".

A Crew-Centered Flight Deck Design Philosophy for High-Speed Civil Transport (HSCT) Aircraft [Palmer et al 1995]

Perhaps the most structured set of design principles relating to operability, these were developed by NASA for future cockpits. They start with three high level philosophy statements, and state a set of principles to support four specified roles the crew play e.g. acting as team members, or being individual operators, which then underpin guidelines on four specific flight deck features e.g. displays, automation. Guidance on conflicts between principles is given. (Material on work activities was being developed separately). The (fairly detailed) principles given for specific features will still need specialist knowledge to generate project-specific design standards.

General principles for the development and use of programmable electronic systems in marine applications [ISO/CD 17894:2001]

Developed by the EC ATOMOS II on Marine Programmable Electronic Systems (PES) [Earthy 1999], ISO 17894 distinguishes between high level product principles (e.g. the PES shall be tolerant of faults and input errors) and life cycle principles (e.g. user centred activities shall be employed throughout the life cycle). In addition, the high level principles are supported by more detailed criteria (e.g. The PES interface should assist the user in avoiding input errors and in detecting input errors where they are made and alert the user when they occur.) and specific guidance for project stakeholders. The traditional difficulties of using high level principles are balanced by specifying a scheme for assessing conformance [Messer 2000] that aims give freedom to the design team without loss of rigour.

Table 4, Examples of principles related to operability

The specification of processes given in ISO PAS 'HS' concentrates on the work activities - even the statements of benefits are related to activity rather than design intent. The decision as to timing and scope of the work programme required to achieve operability as an outcome remains a matter for expertise within the project team. Underpinning ISO PAS 'HS', ISO TR 18529 and ISO 13407 [ISO 13407:1999] is a set of principles that defines human-centred design:

- The active involvement of users and a clear understanding of user and task requirements

- An appropriate allocation of function between users and technology

- The iteration of design solutions

- Multi-disciplinary design.

These principles can be seen as management principles that generate work activities and culture.

The conclusion is that both design principles and work activities need to be specified and their achievement assessed. Palmer *et al* (*op. cit.*) conclude as follows: "*Many engineers and designers would claim that they already perform human-centred design. It is important to note, however, that we do not define human-centred design as simply applying "human factors" to the design process. Rather, we believe that an explicit design philosophy must be clearly described and applied systematically within the framework of a well-defined design process.*" As regards the design process, we would claim that ISO PAS 'HS' meets the criterion of being well-defined, and can be used as the basis for formal assessment. A work programme based on ISO PAS 'HS' would provide the material to enable the achievement of design principles to be assessed.

In summary, we need both design principles and process specifications, but indicators from design principles are lagged, dependent indicators, placing greater weight on the role of process indicators for safety assurance.

4.5 Process and risk management

The effective management of safety risk (= hazard) and project risk (= cost, time) is essential to safety assurance. There are three points to be made about the relationship between risk management and the use of process models.

Firstly, project (or safety) risk assessment can be used as the driver for capability evaluation and subsequent process improvement, i.e. once the risks have been assessed the processes that mitigate the risks can be determined and then used to tailor the assessment and improvement programmes. The MOD uses project risk in this manner for pre-contract award capability evaluations of candidate suppliers [Jones *et al* 1997]. Mandated process improvement is used as a form of risk mitigation.

The second point is that a capability evaluation (possibly relatively quick and informal) is a powerful way of identifying risks (of both varieties) - particularly in relation to management processes.

Finally, the use of standard process models allows organisations such as large customers or regulators to build up cumulative data on generic risks.

4.6 Process and Regulation

To the desk engineer, the regulator appears to have unlimited powers, resources and small print. The reality is somewhat different. There are political limits to regulation and its enforcement, and a real desire to work by encouragement as much as sanction. ISO 17894 is an example of this [Earthy, 1999]. The move to open-textured regulation and broad goal-setting objectives places greater onus on the supply or operating organisation to demonstrate compliance. It is proposed that the process-based approach has much to offer. For example, process improvement can be used as a form of encouragement, but with the potential for formal lawyer-proof evaluations. An informal analysis by one of the authors of (stated or implied) operability process requirements in a range of regulatory settings indicated a high degree of commonality. The prospect of achieving cross-sector commonality by using a process-based approach has considerable attractions.

Some work on process-based approaches to regulation has started. ISO 17894 for the marine sector has a sizeable process element that includes operability. The iCMM developed by the FAA [FAA 1997] has modules that address operability, supported by more detailed guidance [FAA 1999]. The HSE [Blackmore 1996] found that senior line managers had few means to measure health and safety in design, and sought to find indicators of the effectiveness of the design process. It has since sponsored research into the development of Design Process Indicators [Busby *et al* 2000].

In brief, there is work in a number of sectors to exploit the potential of the process-based approach for regulatory purposes. The generic nature of the process models offers the potential for cross-sector commonality.

5 Conclusions and recommendations

The authors conclude that, given suitable support by industry and regulators, a process-based approach has the potential to become the mainstream approach to safety assurance as regards human aspects of systems (and possibly for all aspects, though that goes beyond the scope of the paper). Design principles have been found to complement process measures, but achievement of design principles is dependent upon process capability. In order to adopt a process-centred approach project teams will require detailed method statements for guidance, audit trail and quality purposes, and a 'body of knowledge' to supplement design principles. For practical project purposes, safety culture can be considered separately from a process-based approach, but the use of process metrics can (rightly) relieve safety culture of a role in safety assurance. The possibility of using process metrics in the context of Regulation is intriguing and appears to be under consideration in a number of places.

The principal recommendation is that ISO PAS 'HS', in conjunction with ISO CD 15288, is given further piloting on real projects with an explicit role in safety assurance. There are existing resources for the analysis of trials data that could be used to support this. A related recommendation is to progress the work started by [Benediktsson et al 2001] on the mapping of process models and IEC 61508:1998.

References

[Benediktsson et al 2001] Benediktsson O, Hunter R B, McGettrick A D: Processes for Software in Safety Critical Systems in Software Process: Improvement and Practice, 6 (1): 47-62, John Wiley and Sons Ltd., 2001

[Blackmore 1996] Blackmore G A: Safety Management Systems in Offshore Oil and Gas Companies - Experience from Asessment and Audit of UK North Sea Operations, New Orleans, December 1996

[BS EN 614-1:1995] BS EN 614-1: Safety of machinery - Ergonomic design principles. Part 1. Terminology and general principles 1995

[Busby et al 2000] Busby J, Strutt JE, Sharp JV: "Lessons learnt from Offshore & Marine Incidents & accidents - input to the Design Process", ERA Conference on Major Hazards Offshore, London, November 2000

[Earthy 1998] Earthy J V: Usability Maturity Model: Human-Centredness Scale. IE2016 INUSE Deliverable D5.1.4s, 1998. http://www.lboro.ac.uk/eusc

[Earthy, 1999] Earthy J V: A new approach to marine programmable systems assessment, 9th Intnl. Symp. of the Intnl. Council On Systems Engineering (INCOSE), Brighton, UK, 1999

[Messer, 2000] Messer A.C: Draft Survey Procedures, Assessment scheme realisation, ATOMOS IV ref A408.02.08.055.001, 2000 www.atomos.org/atomos2/

[FAA 1997] Ibrahim L, Deloney R, Gantzer D, LaBruyere L, Laws B, Malpass P, Marciniak J, Reed N, Ridgway R, Scott A, Sheard S: The Federal Aviation Administration Integrated Capability Maturity Model, (FAAiCMM), Version 1.0, 1997 http://www.faa.gov/aio/ProcessEngr/iCMM/

[FAA, 1999] FAA Human Factors Job Aid, Federal Aviation Administration Office of the Chief Scientific and Technical Advisor for Human Factors, AAR-100, (202)267-7125, 1999 www.hf.faa.gov

[HS(G)65: 1991] Successful Health and Safety Management. HSE Books, 1991

[HSG48: 1999] Reducing error and influencing behaviour. HSE Books, 1999

[IEC 61508:1998] IEC 61508: Functional safety of electrical/electronic/programmable electronic safety-related systems, 1998

[ISO/CD 17894:2001] ISO/CD 17894: Ships and marine technology - Computer applications - General principles for the development and use of programmable electronic systems in marine applications, 2001

[ISO 6385:1981]   ISO 6385: Ergonomic principles of the design of work systems, 1981

[ISO 9126:2000]   ISO 9126: Software product quality - quality model, 2000

[ISO 12207:1995]   ISO 12207: Software process - Software lifecycle processes, 1995

[ISO 13407:1999]   ISO 13407: Human-centred design processes for interactive systems, 1999

[ISO CD 15288]   ISO CD 15288: System engineering - System life cycle processes.

[ISO TR 15504:1998-2]   ISO TR 15504: Software process assessment - A reference model for processes and process capability, 1998

[ISO TR 18529:2000]   Ergonomics of human system interaction - Human-centred lifecycle process descriptions, 2000

[ISO PAS tba:2002]   ISO PAS tba:2002 A specification for the process assessment of human-system issues, 2002 *in press*

[Jones *et al* 1997]   Jones R L, Hamilton J M: A co-ordinated approach to identifying software development risks in MOD projects, Proc. European SEPG conference, 1997

[Palmer *et al* 1995]   Palmer MT, Rogers W H, Press H N, Latorella K A, Abbott T S: A Crew-Centered Flight Deck Design Philosophy for High-Speed Civil Transport (HSCT) Aircraft, NASA Langley Research Center, NASA Technical Memorandum 109171, 1995

[Paulk *et al* 1993]   Capability Maturity Model for Software, Version 1.1 CMU/SEI-93-TR-24 Software Engineering Institute, Carnegie Mellon University, Pittsburgh, 1993

[Reason 1990]   Reason J T: Human Error, Cambridge University Press, 1990

[Turner 1978]   Turner B A: Man-Made Disasters. Second edition: Turner B A, Pidgeon N K: Butterworth-Heinemann, 1997

[Vaughan 1996]   Vaughan D: The Challenger Launch Decision. Risky Technology, Culture and Deviance at NASA. The University of Chicago Press, 1996

[Westrum 1997]   Westrum R: Social Factors in Safety-critical Systems. in Redmill F, Rajan J: Human Factors in Safety-critical Systems, Butterworth-Heinemann, 1997

Part of the authors' work was carried out under projects for the European Commission and the UK Ministry of Defence. The support of these bodies is gratefully acknowledged. The opinions expressed in this paper are the authors' own and not those of Lloyd's Register. Parts of this paper have previously appeared as white papers at www.processforusability.co.uk

SAFETY REQUIREMENTS

Human Factors Considerations for System Safety

Dr. Carl Sandom
Thales Defence Information Systems
carl.sandom@uk.thalesgroup.com

Abstract

The author has found that in many cases industry produces Safety
Case reports that provide only limited safety assurance and, like the
proverbial 'head in the sand' ostrich, the dominant system safety risks
associated with the human factors are too often ignored. This paper
provides a brief outline of the Human Factors discipline and its
important relationship with systems safety. The paper then provides a
discussion on some of the more commonly experienced human-
factors problems relating to systems procurement, human-computer
interaction and organisational issues before making some modest
proposals for improvements in these areas. The paper concludes that
the application of Human Factors techniques promotes engineering
solutions that take account of human capabilities and limitations
which can address the major risks to systems safety.

1 Introduction

The aim of this paper is to raise awareness of the need to address the safety risks
inherent with the Human Factors issues relating to the design, operation and
maintenance of safety-critical systems. The paper begins with a brief discussion on
the relationship between Human Factors and systems safety and then provides a
working definition of Human Factors. In the author's experience, some important,
but often neglected, areas of risk relating to Human Factors issues are systems
procurement, human-computer interaction and organisational factors. This paper
will provide a discussion on some of the prevalent problems encountered in these
areas before considering reasons why this situation currently exists and providing
some suggestions for improvement. It should be emphasised that this is not a
taxonomy for all Human Factors-related risks but merely convenient headings to
discuss specific problems encountered in practice. Although many of the views
expressed in this paper are particularly relevant to the Defence Sector, it is believed
they are also appropriate for other industry sectors.

A distinction needs to be made here between different types of risk which exist
in any systems development project. Risks come in many forms and those relevant
to this paper are: risks to a procurement programme and safety risks posed by a

system. It should be noted that contractual issues often raise programme risks; although they can become risks to safety if safety issues are avoided; the reverse is also true for organisational and interaction risks.

The views expressed in this paper are those of the author and do not necessarily reflect the policies of Thales Defence Information Systems.

2   Human Factors

The development of a safe system relies on the integration of many different engineering skills, such as software and hardware engineering, for example. The application of Human Factors techniques is often not well understood by those practicing the more traditional engineering disciplines. This section will outline the importance of Human Factors for systems safety and provide a working definition of Human Factors.

2.1   Human Factors and Safety

This paper deals with some of the Human Factors problems associated with the procurement of complex, interactive systems which can be characterised as systems that support dynamic processes involving large numbers of hardware, software and human elements that interact in many different ways [Perrow 1984]. Typical examples of such systems are found in Air Traffic Control, Ambulance Control Rooms and Power Generation Plants. Some complex systems can be safety-critical and these typically rely on people, procedures and equipment to function safely within an operational environment. For systems such as these, it seems obvious to state that the Human Factors associated with the designers, operators and maintainers must be taken into account when making claims about systems safety.

Despite this, safety cases for complex systems containing human operators often consider safety predominantly, or even exclusively, from a technical perspective. Safety cases such as these are typically limited to addressing the hazards arising through technical failures alone, despite the fact that human error is repeatedly mentioned as a major contributing factor or even the direct cause of many accidents or incidents. For example, an analysis of causal factors contributing to a situation in which the safety of aircraft was compromised show that approximately 98% of incidents in UK airspace during 1997 were caused by human error (calculated from [CAA 1998a] and [CAA 1998b]). Names such as Herald of Free Enterprise, Clapham Junction and Ladbrook Grove are a grim reminder of disasters which have included human failures in complex systems.

Paradoxically, industry too often concentrates the majority of safety assurance effort upon technical issues often neglecting the human contribution. The human component of safety-critical systems are rarely considered to be safety-critical and are not therefore subject to hazard analysis and risk assessment to the same degree as any other safety-critical system component. The conclusion to be drawn from this is that in many instances, industry produces safety cases that, at best, provide only limited safety assurance, as the prevalent errors are related to the Human Factors.

2.2   Human Factors Integration

Human Factors is basically about the need to match technology with humans operating within a particular environment; this requires appropriate job and task design, suitable physical environments and workspaces and human-machine interfaces based upon ergonomic principles. Systems using computers must demonstrate how their human-computer interfaces can foster the safe, efficient and quick transmission of information between the human and the machine, in a form suitable for the task demands. The British Defence Standard 00-25 defines Human Factors as:

" ...*an interdisciplinary science concerned with influencing the design of manned systems, equipment and operational environments so as to promote safe, efficient and reliable total system performance.*" [IDS 00-25/12, p.4, 1989]

Theory is one thing, but practitioners are interested in the pragmatic integration of Human Factors within the systems development life-cycle. Through the application of Human Factors theory and appropriate techniques it is possible to analyse and optimise the human interaction with a system and its environment. A key aim of Human Factors expertise is to minimise safety risks occurring as a result of the system being operated or functioning in a normal or abnormal manner.

Human Factors Integration (HFI) is a phrase used to denote an engineering discipline that applies theory, methods and research findings from ergonomics, psychology, physiology and other disciplines to the design of manned systems. HFI has replaced MANPRINT (MANpower PeRsonnel and INTegration) as the process for managing HFI in defence procurement. It is still structured in broadly the same way, with the six domains of Manpower, Personnel, Training, Human Factors Engineering, System Safety and Health Hazard Assessment as listed in Table 1.

Unlike engineering parameters, Human Factors parameters are not always easily quantifiable. Therefore it is often necessary to use the services of Human Factors practitioners who can use experience to interpret the situation and provide informed predictions when it is not possible to meaningfully measure human performance or relevant criteria.

This brief discussion should give an idea of the type and scope of activities which need to be undertaken by Human Factors professionals and integrated into typical systems development projects. The paper will now look at some of the more common Human Factors-related problems experienced by the author.

HFI DOMAINS	DESCRIPTION
Health Hazard Assessment	Identification and consideration of conditions inherent in the operation or use of a product (e.g. vibration, fumes, radiation, noise, shock, recoil etc) which can cause death, injury, illness, disability or reduce the performance of personnel.
Human Factors Engineering	The comprehensive integration of human characteristics into product design, including all aspects of workstation and workspace design including accommodation / habitability issues.
Manpower	The number of men and women required and available to operate and maintain the product / system.
Personnel	The aptitudes, experience and other human characteristics (including body size & strength) necessary to achieve optimum performance.
System Safety	Application of Human Factors expertise to minimise safety risks occurring as a result of the system being operated or functioning in a normal or abnormal manner. The objective is to minimise to as low a level as reasonably practicable the risk of injury to personnel and damage to equipment.
Training	Specification and evaluation of the optimum combination of instructional systems, education, on job training required to develop the knowledge, skills and attitudes needed by the available personnel to operate and maintain the product to the specified level of effectiveness under the full range of operating conditions.

Table 1: Human Factor Integration Domains

3   System Procurement Issues

The system procurement process is perhaps not usually a topic directly associated with systems safety. However, there is a human factor at work during system procurement and the seeds of failure to adequately address the Human Factors, or even safety, can be sown during the contract negotiations. This section will discuss some of the typical procurement-related problems and risks facing both procurers and potential systems developers.

3.1   Contractual

Generally, a systems procurer, or customer, issues an Invitation to Tender (ITT) for a development contract to a number of competitive contractors. In response to an ITT, the invited contractors must then put together proposals which are both technically and financially persuasive. In today's highly competitive markets, the financial arguments are often more compelling to the potential customer. This inevitably means that all engineering costs within a proposal must be defensible and as such must be perceived to add value to the product.

Customers will inevitably have financial constraints and will, naturally, aim to procure maximum functionality for the minimum cost. Of course the problem here is that the output from Human Factors or safety analyses are generally non-functional requirements which constrain the design of a system rather than add functionality. Consequently, Human Factors or Safety Engineering activities can be perceived to be unnecessary, or unwanted, additions to a contract and the temptation when considering their associated costs is to limit their impact on the contract, or worse still, to ignore them.

From the contractor's perspective, a response to an ITT must contain a proposal which is competitive in every respect in order to win business. With such financial pressure it is difficult, if not impossible, to compete effectively if the cost of Human Factors analyses are added without them having been explicitly requested by a customer. If the contractor believes that there may be a Human Factors-related risk to safety, what could happen is to explicitly exclude such activities from their bid to limit their financial risk, or worse still, to ignore them.

Thus, there can be a situation where both parties are either genuinely unaware of the safety implication of avoiding Human Factors activities or they may individually choose to act like the proverbial ostrich and, metaphorically, stick their heads in the sand and ignore it.

To address these potential seeds of failure, both customers and suppliers need to have a shared appreciation of the value added to a product by Human Factors analyses and the vital link with systems safety. In complex, safety-critical systems there must be a clear customer requirement for safety assurance activities to be underpinned with Human Factors analyses. This would ensure that all contractors cost for the safety-related Human Factors activities from the outset of the project enabling safety to be designed into the product rather than the unsatisfactory alternative of mitigating risk with relatively soft procedures at a later date.

3.2   Operational Environment

An alternative to the picture painted so far is that both parties agree that Human Factors analyses are required from the outset. As with any other system requirement, Human Factors requirements need to be specified in the user and system requirements documents. In practice, this can be difficult as a analysis of Human Factors in safety-critical systems often reveals a complex set of problems relating to the people, procedures and equipment (or technology) interacting within a *specified* environment as depicted in Figure 1.

The environment referred to here consists of both the immediate working environment and the wider operational environment. The working environment raises many physical, ergonomic issues such as lighting, noise or heating levels and the effects upon the human operator in terms of stress, attentiveness, etc. Generally, the working environment for a system is easily specified and is not considered further here. However, the operational environment for a system is typically much more difficult to specify, as it requires precise knowledge of how the system will be used and this must be documented in detail. This doesn't just mean specifying the *interfaces* to a system (for example, a communications link protocol at the functional level); this also means specifying the tactical and strategic uses of the system (for example, how calls are prioritised in an emergency services control system).

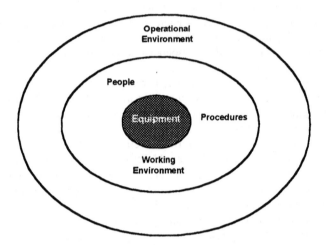

Figure 1: System Boundaries

The specification of the operational environment is important from a Human Factors perspective as this is typically used, along with the people and equipment issues, to develop an *operational concept* and from this, specific operational procedures to ensure that a system is used safely and efficiently. A systems developer would also use the operational concept as the basis for developing the system requirements specification upon which the Human Factors analyses are founded. The system safety analyses, including Human Factors task analyses, must be based upon a comprehensive system requirements specification to ensure that all credible hazards are identified.

In practice, it is difficult, if not impossible, to completely specify the operational concept of a system at the outset and a user may only specify the functionality required of a system without having a complete appreciation of how that functionality will be used in the operational environment. Perhaps this should not be unexpected and there are many explanations of unforeseen environmental changes introduced with the adoption of new technology (for example see [Macredie and Sandom 1999]). The specification of the operational environment can also be a problem when a business sector is dealing with national security and the users are reluctant to make the operational aspects explicit for either new or replacement systems. However, the outcome in both cases can be a system that implements a 'bag of functions' without taking account of the wider operational aspects.

So what can be done about this? Ideally, before an ITT is sent to potential contractors, the system procurer would ensure that the users have decided exactly how the system will be used to permit the specification of the operational environment. This may sound obvious, but in practice it is not unknown for systems to be delivered without the operators fully specifying the intended system use. If the operational concept cannot be fully specified from the outset, for whatever reason, it is necessary to make a number of explicit assumptions concerning the operational issues. It is essential that these assumptions are ratified by all system stakeholders and documented to mitigate the financial risk of rework due to changing operational procedures as the systems development proceeds. Finally, both systems procurers and developers must recognise that any changes to the operational concept or assumptions will impact upon the Human Factors analyses in particular.

4   Human Computer Interaction

Another area where problems are common and misunderstanding is rife concerns system human-computer interactions. To those unfamiliar with the Human Factors discipline, it can be the case that Human Factors are associated entirely with the analysis of the human-computer interface. As discussed previously, this is not so, however, the interface is certainly an important aspect. The term Human Computer Interaction (HCI) is used here to mean the *process* of communication between users and a system rather than simply the implementation of the interface (ie. not just human-computer interface).This section will discuss two common problem areas associated with the assessment of human computer interactions in safety-critical systems; namely human reliability and usability.

4.1   Human Reliability

HCI issues are important for safety-critical systems. For complex systems in dynamic environments, an operator must pay attention to a large volume of information from a variety of sources, including sensors and other operators, in order to acquire an awareness of the situation in question. In many cases humans are no longer able to appreciate the true situation without the aid of machines, therefore, machines must tell us more of what we need to know and they must do it more effectively and less ambiguously than before [Billings 1995]. The quality of the information acquired through the interface can contribute significantly to human failure, and the design of the human-computer interface can have a profound effect on operator situational awareness and system safety. When emergencies arise and system operators must react quickly and accurately, the usability of the system is critical to operators' ability to make decisions, revise plans and to act purposefully to correct the abnormal situation. Analyses of human failures in large control centres have repeatedly shown that operator errors are linked with poor control layout and misleading cues [Booher 1990].

Systems safety assessments are predicated upon calculations of the inherent *dangerous* failure rates which are typically a sub-set of all failures and are therefore not a measure of system reliability. As discussed previously, human failures are typically the most prevalent in a system; yet they are often overlooked by system developers. This may be because hardware reliability techniques are relatively mature and well understood, however, this is not the case when dealing with human reliability. It is very difficult, if not impossible, to predict all the potential mental states of an operator in a complex system. Even if it were possible to identify all the potential mental states, and their effects on human behaviour, the difficulty of estimating the probability of occurrence of each state remains. Human Reliability Analysis (HRA) techniques have attempted to address this issue [see Kirwan 1994].

Arguably however, to a large extent the quantitative aspects of HRA research have been dominated by assumptions that apply to technical systems and often these do not translate to human systems [Woods *et. al.* 1994]. The hazards associated with human failures are very different from the hazards which have historically concerned system designers since they arise directly from the *use* of the system and therefore require some understanding of the cognition and actions of users within the operational environment. This aspect is critical to systems safety assessment, yet it often does not fit with 'conventional' views on systems engineering practice.

4.2   Usability

Usability is another popular term used in system specifications - yet there is no accepted definition. Nonetheless, usability is generally taken to mean not only ease of use but the concept also equally involves effectiveness in terms of measures of human performance [Smith 1997]. From this general definition, safety-critical system developers may be tempted to infer that a useable system is, by implication, a safe system. However, usability and safety can be mutually exclusive system properties. It is possible that making an interactive system safe will entail many trade-offs with usability. For example, interface prototyping may reveal a requirement for a complex keying sequence to be replaced with a macro facility allowing a function to be invoked with a single key press. This requirement may enhance system usability, however, it may inadvertently affect the safety of the system if a hazard is associated with the function being invoked. This point may seen obvious, but systems operators and others involved in HCI Working Groups will often support system usability without being aware of all the safety issues, and these views often prevail in the design.

While a complex sequence may not be very efficient in terms of usability, it provides a number of opportunities for the operator to become aware that the function being invoked may be hazardous in the current context. It is suggested here that the greatest hazard in a system can be associated with an operator automatically interacting when conscious thought is required. With familiarity, automatic human cognition can become the norm and information is then perceived, interpreted and acted upon with little or no thought. Conscious cognition bears a complex relationship to situational awareness, and it seems intuitively unsafe to perform safety-critical tasks while remaining unaware of them even if they are performed well [Hopkin 1995]. The implication is that operator awareness of a situation may not be updated and may therefore be inaccurate. It can be concluded that it is not enough to simply concentrate on the usability of an interactive system to assure functionally safe operation.

4.3 HCI Safety Assessment

Given the difficulties outlined here relating to the assessment and mitigation of Human Factors risks, it may be argued that human error is best examined from a cognitive perspective, as traditional reliability engineering techniques do not appear to fit well with Human Factors concerns. It has been suggested that safety and usability can be mutually exclusive properties, particularly in systems that rely on situational awareness for safe operation. If this is the case, different methods and techniques are required for evaluating safety. It may be more appropriate to quantify safety from a Human Factors perspective in terms of the level of situational awareness acquired through the interface.

A complete discussion on situational awareness is beyond the scope of this paper (see [Sandom 1999] for a detailed discussion). Briefly, however, operator situational awareness can be considered as a mental state acquired through a process of interaction. To assess the impact of situational awareness on systems safety, it is equally important to assess both the mental state and the process. Situation Awareness Global Assessment Technique (SAGAT) is a popular method of assessing the mental state of an operator [Endsley 1995] and Situation Awareness Process Analysis Technique (SAPAT) [Sandom 2001] can be used to assess the acquisition process.

SAPAT in particular aims to identify those potentially hazardous interactions and can help systems developers to make informed trade-offs between usability and safety.

5 Organisational Issues

There are many organisational factors at work during the development of any complex system. This section will discuss some of the typical problems facing systems developers that try to adopt an integrated approach to Human Factors.

5.1 Organisational Failures

Most people would recognise that all systems have a human input, even if it is limited to the fallibilities of the developers who can introduce systematic errors into the design and implementation phases. However, it is perhaps not always appreciated that organisational issues can profoundly affect systems safety.

Consider a simple example of a hypothetical system (adapted from [Woods et. al. 1994]). This system has an operator who is required to enter a number when a screen flashes A and enter another number when a screen flashes B. If an operator one day enters the A number when B appears on the screen and the system blows up instead of shutting down, some would conclude that the accident was caused by human error and that would be the end of the investigation. However, that wouldn't help us to understand anything at all about cause and effect. Considering issues such as the system design, and understanding how operators solve problems of workload and competitions among goals, would provide a more meaningful investigation. Moreover, in the simple example, it could be argued that the failures had been made by the organisation, which is to say people such as designers and managers who created the poor conditions for the operator error.

To address the root cause of the organisational risk to human error it is necessary for the organisation to develop a positive safety culture based upon a sound safety management system. This is often more difficult than it may appear and it cannot be achieved overnight. As for most organisational changes, a safety culture must be adopted from the top down, and senior management may need to be educated to understand that a safety culture cannot be contracted into an organisation via outsourcing agreements. Another common misconception is that an organisation is 'doing' safety if it has an ILS policy in place. Consider how many organisations equate safety with reliability - yet a system may reliably perform a specified function which is unsafe.

5.2   Safety Culture

As manufacturers and suppliers of goods and services, companies have a responsibility and must be fully committed to a policy of compliance with product safety legislation. Product safety should assume prime importance in the design, development, manufacture, assembly, operation, support, maintenance and disposal of company-designed products. Yet it is too easy to focus on the specified *contractual* safety requirements without appreciating that there are legal requirements that must also be addressed. Clearly these obligations will overlap to different degrees, as shown in Figure 2, but an organisation must address both obligations nonetheless.

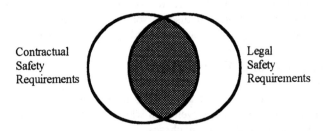

Contractual
Safety
Requirements

Legal
Safety
Requirements

Figure 2: Safety Obligations

A problem can occur for organisations when a contract does not explicitly require Human Factors analyses to be undertaken. Given the previous discussions on the impact of Human Factors issues on safety, it could be argued that the organisation has a legal duty of care to ensure that any related risks are reduced to a level as low as reasonably practicable through Human Factors analyses, regardless of contractual obligations. In short, organisations that profess to have a safety culture should also have Human Factors expertise.

5.3   Selling Human Factors

Throughout this paper it has been argued that Human Factors is an important element in the design of effective and safe systems. However, before any profit-making organisation can be expected to fund a Human Factors capability, a sound business case must be made to show the potential financial benefits. In practice, this is not easy, as the costs and benefits associated with typical Human Factors activities are often difficult to quantify.

A detailed explanation of making a Human Factors business case can be found in [Trenner and Bawa 1998]. However, to summarise, early integration of Human Factors in the design and system life-cycle promotes solutions that take account of human capabilities and limitations. The safety-related benefits include: enhanced usability, reduced error rates, and improved in-service performance. Also, early integration of Human Factors into the design process helps reduce the number of design changes and associated costs throughout the whole product life-cycle.

6   Providing Safety Assurance

A crucial question that may have been at the back of the readers mind: What can be done about the concerns discussed here and the apparent lack of safety assurance caused by the neglect of Human Factors? After having looked at some of the common problems, what are the solutions? Perhaps disappointingly, this paper does not claim to provide all the answers but merely makes some modest proposals. However, this section will demonstrate how a Human Factors argument could be made for a system safety case.

6.1   Making an HF Argument

A System Safety Case will typically contain arguments and supporting evidence that the system meets or exceeds the required standard of safety. Broadly, the arguments must show that the risks associated with operating or maintaining the system have been reduced to a tolerable level. A main safety objective is to validate all safety requirements and show that they have been successfully implemented. The prevalence of human failures in complex, safety-critical systems has already been discussed. If we accept that Human Factors can contribute significantly to the safety risks in these systems, then a safety argument must explicitly address these issues.

In the construction of safety cases there is a large amount of information to be recorded and managed. Goal Structuring Notation (GSN) is one notation that has been developed to allow hierarchical structuring of such information and is used to express high level arguments with links to supporting evidence. A subset of GSN will be used here to illustrate a Human Factors safety argument.

Briefly, GSN uses a number of concepts including Goals, Strategies and Contextual Information. A Goal can be considered as a statement of a requirement to be met by a system, or some activity to be performed, while a Strategy introduces an element of explanation showing how the safety arguments are constructed. Finally, Contextual information is often necessary to understand a goal or strategy. Figure 3 shows a Human Factors safety argument expressed in GSN.

The safety argument in Figure 3 is based upon the following example, high-level Human Factors requirement:

System Requirement [HF1]: *"The design and implementation of System X shall include consideration of Human Factors Integration issues"*.

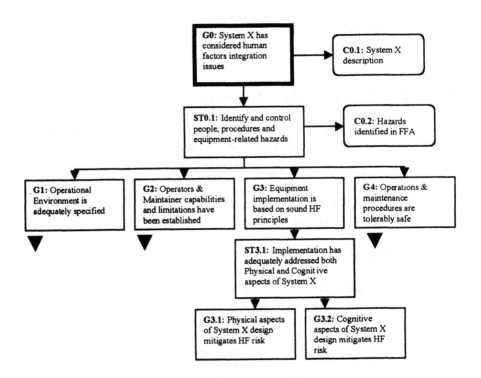

Figure 3: Example Human Factors Safety Argument

Figure 3 shows the top-level goal G0 as a statement of the requirement HF1 to be met by a system developer. In this example, the goal G0 uses the contextual information C0.1 which provides a system description. It has already been stated that safety-critical systems typically rely on people, procedures and equipment to function safely within an operational environment. For systems such as these, it seems logical to use a strategy like ST0.1 in Figure 3 (supported by the contextual information of the safety analyses like the FFA) to structure a Human Factors safety argument in these terms.

Each specific sub-goal in this argument would typically require direct supporting evidence or further decomposition as shown by Goal G3. The type of evidence would be determined by the assurance level required from the system and should ideally include both formative and summative evidence to underpin a Human Factors argument. This approach would be consistent with Human Factors methods which are broadly categorised as formative, generating evidence during the overall system design process, and 'summative', generating evidence relating to the evaluation of the final product (Noyes and Baber 1999).

Figure 3 shows only the top-levels of a safety argument and has only decomposed Goal G3 to illustrate the concept. In this example, to follow the strategy of ST3.1, Goals G3.1 and G3.2 would broadly deal with the ergonomic and cognitive aspects of the system respectively. So G3.1 might rely on anthropometric evaluations, for example, while G3.2 would perhaps use evidence generated by SAPAT analyses or HCI Working Groups.

It is not claimed here that this simplistic and partial example of a Human Factors safety argument is the only way to structure an argument. It is recognised that there are many different ways that an argument can be expressed. However, the example in Figure 3 does illustrate one way in which systems developers could start to address the majority of the risks associated with the operation of complex, safety-critical systems.

7   Conclusions

This paper is intended for an audience of both engineering practitioners and academics, to raise awareness of the need to address the safety risks inherent in Human Factors issues relating to safety-critical systems. The important relationship between human factors and systems safety was discussed and it was suggested that human factors issues typically represent the biggest safety risks in complex systems. A working definition of Human Factors was given as influencing the design of manned systems, equipment and operational environments to promote safe, efficient and reliable total system performance. Some important Human Factors considerations relating to systems procurement, Human-Computer Interaction and organisational factors were discussed and some suggestions for improvement were provided. Finally, an example of a Human Factors safety argument was provided to illustrate how systems developers might start to address the majority of the system safety risks and to provide holistic systems safety cases.

From these discussions it can be concluded that Human Factors should be an important consideration in the design of effective and safe systems. Early integration of human factors in the design and system life-cycle promotes solutions that take account of human capabilities and limitations. The safety-related benefits include: enhanced usability, reduced error rates and improved in-service performance. Like safety, human factors is not like a coat of paint that can be applied at the end of a project, and addressing some of the issues highlighted in this paper may help industry to produce systems that provide the necessary level of safety assurance.

7 References

[Billings 1995]   Billings C. E: Situation Awareness Measurement and Analysis: A Commentary, in Garland D. J. and Endsley M. R., (Eds), Experimental Analysis and Measurement of Situation Awareness, Proc. of an Int Conf, FL:USA, November 1995

[Booher 1990] Booher H R (Ed.): MANPRINT – An Approach to Systems Integration, Van Nostrand Reinhold, 1990

[CAA, 1998a]   Civil Aviation Authority: Aircraft Proximity Reports: Airprox (C) - Controller Reported, August 1997 - December 1997, Vol 13, Civil Aviation Authority, London, March 1998

[CAA, 1998b] Civil Aviation Authority: Analysis of Airprox (P) in the UK: Join Airprox Working Group Report No. 3/97, September 1997 - December 1997, Civil Aviation Authority, London, August 1998

[Endsley 1995] Endsley M R: Measurement of Situation Awareness in Dynamic Systems, Human Factors, 37(1), 65-84, March 1995

[Hopkin 1995] Hopkin V D: Human Factors in Air Traffic Control, Taylor and Francis, London, 1995

[IDS 00-25/12 1989]   UK Ministry of Defence Interim Defence Standard 00-25 (Part 12)/Issue 1, Human Factors for Designers of Equipment, Part 12: Systems, July 1989

[Kirwan 1994] Kirwan B: A Guide to Practical Human Reliability Assessment, Taylor and Francis, London 1994

[Macredie and Sandom 1999]   Macredie R D and Sandom C: IT-enabled Change: Evaluating an Improvisational Perspective. European Journal of Information System, 8:247-259, 1999

[Noyes and Baber 1999] Noyes J and Baber C: User-Centred Design of Systems, Springer-Verlag, Berlin, 1999

[Perrow 1984] Perrow C: Normal Accidents, Princeton Unuiversuty Press, 1984 (2nd edition, 1999)

[Sandom 1999] Sandom C: Situational Awareness through the Interface: Evaluating Safety in Safety Critical Control Systems. IEE Proceedings of People in Control, University of Bath, UK, 21 – 23 June 1999

[Sandom 2001] Sandom C: Situational Awareness, in Noyes J and Bransby M (Eds.), People in Control: Human Factors in Control Room Design, IEE Publishing, November 2001

[Smith 1997]   Smith A: Human-Computer Factors: A Study of Users and Information Systems, London, McGraw Hill, 1997

[Trenner and Bawa 1998]   Trenner L and Bawa J: The Politics of Usability, Springer-Verlag, Berlin, 1998

[Woods *et. al.* 1994]   Woods D D, Johannesen L J, Cook R I and Sarter N B: Behind Human Error: Cognitive Systems, Computers and Hindsight, CSERIAC SOAR 94-01, Ohio State University, December 1994

Will it be Safe? - An Approach to Engineering Safety Requirements

Alan Simpson and Joanne Stoker

Praxis Critical Systems Limited

20 Manvers Street, Bath BA1 1PX, UK

Email: alan.simpson, joanne.stoker @praxis-cs.co.uk

Abstract

This paper describes experiences using Safety Requirements Engineering (SRE) to reduce the risk of systems not achieving safety certification and not working safely in the intended environment. Industry is creating ever-larger systems with increasing complexity. Applying traditional process-based safety assurance has become unwieldy and uneconomic. In this paper we describe some practical techniques we use for SRE to support rigorous product-based assurance. The aim is to provide detailed safety requirements to guide the choice of design, architecture and development processes, thus managing assurance more effectively. We describe an overview of the approach with specific emphasis in the second part of the paper on accident and hazard identification, illustrated with some examples from previous projects.

1 Introduction

It is well documented, e.g. [HSE 1995], that a significant proportion of accidents are caused by mistakes in requirements. A significant number are attributed to human error. Whilst the latter were not all caused by mistakes in requirements, we should at least understand the relationship between humans and their environment, and the requirements imposed on them. Investigations into the causes of accidents, such as Nancy Leveson's report [Leveson 2001] or the sources of software error, such as [Lutz 1993] indicate that more rigour is required in setting the requirements and specification of safety related systems.

This paper presents an approach for identifying safety requirements and the specification of system safety, and provides examples of where parts of the approach have been used in real projects.

The ultimate aim of the Safety Engineering process is to demonstrate that all risks posed by the system are acceptable. This can be achieved through the construction of safety arguments and supporting evidence. One of the first claims that a safety argument needs to make is that the safety requirements are complete and correct. No matter how good the system design, it is only as complete and correct as its requirements. A reliable system is not necessarily safe; in fact it can be reliably unsafe. For example, in an assessment of a safety-critical avionics protection system, it was discovered that one hazard, associated with the timing of outputs, had not been considered. It was argued that this did not matter as the software that monitored the output function was "SIL4". However, the software was not specified to check the timing of the output, therefore, the hazard was unmitigated even though the software was developed to safety-critical standards.

Praxis Critical Systems undertakes safety engineering activities across a number of industrial sectors including Aerospace, Defence, Transport and Medical, in all phases of the safety lifecycle. Our experience from involvement in the development of safety-critical systems, including the preparation of safety requirements and safety arguments, has helped us to develop the way completeness and correctness in requirements is demonstrated.

Early identification of complete, correct, and usable requirements is a pre-requisite for successful project delivery [Vickers 1996]. Yet problems remain and failings in requirements engineering are still one of the principal causes for over-run, over-budget, under-achieving projects and instrumental in some major accidents [Leveson 2001].

In our experience, there are four fundamental problems to be addressed by the safety requirements engineering activity.

- **Absence**: Without coherent safety requirements we do not know what we need to demonstrate. Additionally the system design and development are not constrained to conform to accepted "good practice".
- **Expression**: People often find it hard to articulate their requirements. There is some confusion about what is a safety requirement and what is not. Safety requirements are often developed without adequate domain knowledge.
- **Conflict**: When requirements are identified, they will often conflict. Safety requirements frequently come into direct conflict with other World requirements.
- **Change**: Even if requirements conflict is resolved, the requirements are likely to change. Once identified, safety requirements tend to be quite stable[1], however, the affects of other changes, such as integrating new equipment or reuse in a new environment, need to be considered against the safety requirements.

Praxis Critical Systems developed a requirements engineering method called REVEAL®[2] [REVEAL] for the capture, analysis and documentation of system requirements in order to reduce the risk of systems not working in their intended environments. Whilst the approach does not explicitly address safety requirements REVEAL provides a number of sound principles that add clarity and rigour to the safety requirements engineering process. This paper describes an overview of an

---

[1] Often only because the safety requirements are derived after the design.

[2] REVEAL is a registered trademark of Praxis Critical Systems Limited.

approach to engineering safety requirements using the principles and terminology of REVEAL. The main objectives of the approach are to:

- make safety requirements derivation as rigorous as the rest of the safety-related development[3];
- identify a method that provides better assurance of completeness and correctness in the identification of accidents and hazards;
- realise the benefits of the REVEAL approach to give a precise specification of what a system must achieve in order to work safely;
- provide an auditable and accessible representation that can be discussed and contributed to by all stakeholders.

2 Requirements Engineering using REVEAL

REVEAL is a true engineering method since it is the systematic application of scientific principles, based on those published by Michael Jackson [Jackson 1995], which are augmented with practical engineering experience to provide a rigorous process, recognising that:

- requirements are about the World and need a systematic elicitation process to identify and document what the users need as well as the environment within which the system must operate;
- many people have an interest in the provision of the system and so requirements will conflict[4];
- requirements will change.

For some years the rapid development of technology, particularly computers and telecommunications, has led to systems and projects of increasing size and complexity. In our experience Requirements Engineering (RE) is being increasingly applied to systems that are built by integrating complex subsystems, i.e. "systems of systems". The relationship between the subsystems' requirements and the ultimate goals of the integrated whole are typically complex. Undesired interactions between subsystems can also be very difficult to identify. Consequently, systems integration is a major, or even dominant, risk in the production of systems such as an aircraft, railway or telecommunications infrastructure.

REVEAL uses the Jackson concepts [Jackson 1995] of the World, the Machine, and their Interface to define Requirements (R) and Specifications (S), and uses the term 'Application Domain' to denote those parts of the World we are interested in. This is illustrated in Figure 1 below.

---

[3] This is not about how requirements can be formally specified but how you derive them in the first place.
[4] The REVEAL process actively identifies conflicts and provides guidelines for analysing and resolving conflicts to derive a conflict-free specification

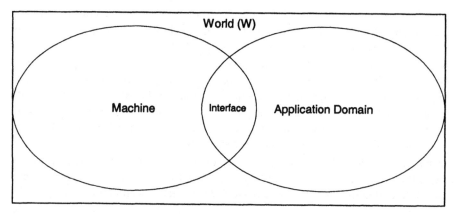

Figure 1 - REVEAL Concepts

The philosophy behind these concepts is that we want to bring about change to some part of the World, and thus we build the Machine to bring about these improvements; the Machine must have some Interface with the World. The method distinguishes between two kinds of statements that we can make about phenomena in the World:

- facts or indicative statements of domain knowledge D that describe properties of the World which are (or are assumed to be) true;
- wishes or optative statements that describe the properties we want to be true - our requirements R.

Requirements (R) are defined as a collection of statements about things in the World (W) that we want the Machine (M) to make true. A Specification (S) describes the Machine's external behaviour, i.e. it includes only shared phenomena in the interface (I) and defines only things the Machine itself can control, whereas a Design is a statement about the Machine itself and, thus, it describes things in M. Note that specifications are sometimes also requirements. They do talk about things in the World, since phenomena in the interface are in both the World and the Machine. Many specification statements are not requirements, though, because they are not in themselves required.

A requirement is said to be satisfied if the specification is met by the Machine in the context of the domain knowledge. This can be represented as a satisfaction argument.

$$D, S \vdash R$$

The Satisfaction Argument should be read as follows: using the relevant properties of the World (D), when combined with the specification of the behaviour of the Machine to be constructed (S), it is possible to show that the Requirements (R) will hold. The Satisfaction Argument therefore means that the specification of the Machine is correct if, and only if, we can show that the specified behaviour of the

Machine and the properties of the domain into which the Machine is put, are together sufficient to achieve the requirements. Within REVEAL the Satisfaction Argument provides a framework for the whole process.

For example, the Application Domain could be Air Travel, i.e. all parts of the World related to air travel, airports, aircraft, passengers etc. A sub-set of the requirements for passenger air travel relate to ensuring the safe passage of aircraft from airport to airport. Therefore, we can use application domain knowledge (e.g. aircraft performance, airport layout, etc.) to define an Air Traffic Management Machine to bring about a reduction in the safety risks.

In the next section we examine the safety engineering process leading up to the derivation of the machine specification from Risk Classification, through Accident and Hazard identification, to the derivation of the safety specification.

3 Engineering Safety Requirements

REVEAL provides a logical framework for the capture and organisation of all requirements in a robust manner. The REVEAL approach does not provide a method for deriving safety requirements but the principles outlined above give us a valuable perspective. Our objective is to demonstrate the completeness and correctness of the safety specifications by deriving a Satisfaction Argument for all of our safety requirements. To do this we need to map the safety engineering concepts behind the identification and specification of safety requirements with the REVEAL concepts.

Safety engineers are interested in assuring the avoidance of accidents. Clearly, accidents only happen in the World, but not all accidents are within the scope of our Machine. Thus, the primary requirement is to prevent accidents, within the Application Domain of the Machine. Therefore, Safety Requirements are the collection of statements about the avoidance of accidents that we want our Machine to make true. The Safety Specification will include all the things that we would like our safety-related[5] Machine to do in order to prevent accidents happening in the World.

However, we will need domain knowledge in order to show that the specification satisfies the requirement. Whilst we would like to specify a Machine such that risk is reduced to nothing, there will always be an element of risk and therefore we need to define acceptability criteria for the risk of an accident occurring, in order to define the amount of risk reduction required.

Whilst the Machine may introduce risk as a consequence of the chosen design implementation or intended functionality, the primary objective for the safety-related Machine is to reduce the risk of accidents. We use domain knowledge to help define the proportion of the risk reduction requirement that the Machine is specified to achieve.

---

[5] In this paper the term safety-related includes safety-critical and safety-involved.

3.1 Risk Classification

Risk is defined to encompass both the likelihood and the consequence of an event so that we may rank the relative risks from different events. There is a variety of methods for classifying risks, but most, such as [Def Stan 00-56], centre on the use of classification tables. Some rely solely on the consequence (ignoring likelihood) of an event whilst others rely solely on the likelihood. This approach assumes that we have only some vague notion about the risks we wish to reduce. But we need to know precisely what the risks within the application domain of the Machine are in order to demonstrate that we have reduced the risks to an acceptable level. Therefore, we should be able to precisely specify the acceptance criteria for each risk.

Many risk tables use just four classifications of severity with very crude distinctions between them, for example *catastrophic* is death and *critical* is serious injury. This can result in the same consequence classification for an accident that can cause one death and one that can cause one hundred, thus the acceptance criteria are the same. Some classifications define *catastrophic* as multiple death and *critical* as single death, but again this is highly subjective and could lead to under classification.

An alternative approach, which is by no means new, is to classify the risk using the accident sequences and the acceptable occurrence rate. For example, rather than define "passenger falls from train door" and "train hits other train" as *catastrophic*, they can be defined individually with intolerable occurrence rate targets[6] as:

- Passenger falls from train must be less than once per year across the UK;
- Train hits other train must be less than once every ten years across the UK.

Accident sequence analysis and casual analysis can be used to apportion these targets down to individual components of the Machine. However, we have to assure that all of the accidents which can occur within the application domain have been identified so that:

- a complete and correct set of safety requirements can be derived;
- the risk apportionment can be carried out correctly.

Although the risk target is apportioned based on the tolerable likelihood of an accident, Machines can still be specified to reduce risk through reducing the severity. Examples of this include automobile air bags and crashworthiness in aircraft and trains. The risk apportionment, therefore, should model the relationship between the tolerable likelihood of the hazard and the accident consequences.

Tolerable risk should also be reduced As Low As Reasonable Practicable. Although this is not discussed further here it is worth noting that only some

---

[6] By definition [HSE 1999] a lower occurrence rate would be tolerable, but should still be reduced As Low As Reasonable Practicable (ALARP). A negligible rate may also be defined below which ALARP need not be applied.

standards (e.g. The Railtrack Safety Management Standard [Yellow Book]) provide guidance on judging ALARP.

In the following sections we outline our engineering safety requirements approach, which is intended to provide a greater degree of rigour in the process of risk identification through to requirement setting.

3.2 Accident Identification

At the beginning of section 3 we argue that the key requirement for safety is the avoidance of accidents. We define the acceptable risk parameters from our knowledge of the application domain to help us complete the satisfaction of the requirement by a Machine. At the highest level a Machine may be specified to reduce the risk of:

- specific known accidents, e.g. we want to reduce the risk of train collision so we specify a machine to control train movements;
- accident scenarios introduced because of the implementation solution chosen for the Machine, e.g. the specification calls for the provision of electrical power. The choice of a nuclear power plant would introduce additional potential accident scenarios;
- accident scenarios introduced because of the nature of the machine's functionality, e.g. a weapons system;
- accident scenarios introduced because of novelty of purpose, e.g. passenger space flights to Mars.

In all cases the Machine must be specified so as to reduce the risk of each accident to an acceptable level. Risk reduction is discussed in section 3.4. Therefore, we need to identify a complete and correct set of accidents within the application domain. Clearly there are accidents we can predict without a specific Machine, but there will always be the possibility of accidents because of the Machine. To assure completeness and correctness of the Machine safety specification we must consider both an "accident-down" view and an "implementation-up" view.

Our approach to the identification of accidents is discussed further in section 4.

3.3 Getting from Accidents to Machine Specifications

The Machine specification needs to include statements common to both the machine and the World. Therefore, we need to identify the contribution of the Machine to accidents. If we consider activities in the World as a series of state transitions then our primary safety requirement is the avoidance of accident states. We are not necessarily concerned with the identification of all states or transition events, but rather with the particular states of the Machine from which accidents can occur thus leading to an accident state.

[Def Stan 00-56] defines a hazard as: *"A physical situation, often following from some initiating event that can lead to an accident"*. Accidents occur when the World is in a hazard state and all necessary conditions exist to give rise to an accident state. This in turn can trigger other accidents to happen and other accident states to be entered. There are a number of reasons why it is important to view a hazard as a state in the World:

- state includes not only the concept of event but also time at risk;
- the set of events that can lead to a hazard is not just a sub-set of Machine failures;
- there are many more events that can lead to a hazard than there are hazards[7].

Not all hazards are relevant to the Machine so we need to define the context of the Machine. In REVEAL this is called the Problem Context or the Application Domain, which is those entities in the World (people, systems, equipment, hardware, etc.) that are relevant to the problem, together with the boundary of the Machine. Thus we define three classes of hazards as illustrated in Figure 2 below.

- World Hazards: those outside the Application Domain.
- Domain Hazards: those within the Application Domain.
- Machine Hazards: any states of the shared phenomena that are Domain Hazards or give rise to Domain Hazards.

A Machine Hazard may not manifest itself as a hazard in the World directly, for example, a protection system failure may not cause a Domain Hazard, but instead increases the risk that the Domain Hazard will occur.

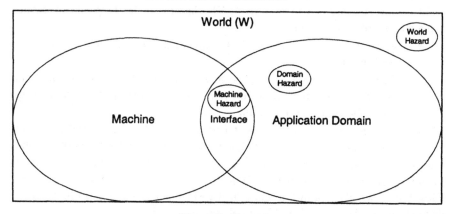

Figure 2 –Hazards

In order to develop the safety specification for the Machine we need to identify all of the Domain Hazards and Machine Hazards. States within the Machine may be defined as hazards, but we are only interested in this classification if we want to define hazards for sub-Machines. We would do this where we need to construct a specification for a component of the Machine. To define a sub-Machine we would redraw Figure 2 with the sub-Machine as the Machine. The Application Domain of

---

[7] When we identify a hazard that can arise from an identified hazardous event then we can ask what else can cause the hazard.

the sub-Machine includes the parent Machine and any other Machine that relies on the sub-Machine, in that the requirements of the sub-Machine are derived from:

- the design of the parent Machine(s);
- the interaction with other sub-Machines, and;
- any interaction with the Application Domain of the parent Machine(s).

For example, an Aircraft comprises many sub-systems including engine, avionics, landing gear, etc. The engine is specified to, amongst other things, provide sufficient thrust for the aircraft to achieve flight, to provide power to other sub-systems and to keep noise to a minimum.

The aim of the accident and hazard analysis at the requirements stage is to model the causes of each accident down to the level at which we can specify safety statements for the Machine, i.e. down to the Machine Hazards. For a typical safety-related Machine, there may be many possible scenarios, or chains of events leading to the accidents. The chain of events forming each scenario would be of the form:

$$\text{Machine Error(s)} \rightarrow \text{Machine Hazard}^8 \rightarrow \text{Domain Hazard(s)} \rightarrow \text{Accident(s)}$$

We refer to these chains as hazardous event chains. For completeness, the hazard identification process needs to model all of the hazardous event chains from accident to error source. Our approach to identifying and modelling the hazardous event chains is described in section 4. The hazard model helps us to capture domain knowledge so that we can:

- refine the requirements derived from identified accidents down to the Machine hazards;
- decompose what we want to be true in the World (i.e. risk of accidents reduced to an acceptable level) into what we want the Machine to make true in the application domain;
- apportion the accident risk target to each Machine hazard, i.e. the proportion of the overall risk budget that must be reduced by the Machine.

The Machine hazards are the point at which the Machine can contribute to hazardous events in the application domain. It is also the point at which hazard mitigation is wholly within the scope of the Machine. Machine hazard mitigation therefore forms the basis of the Machine's safety specification.

The specification of the Machine needs to completely and correctly define the mitigation of each hazard and must include statements, which when combined with domain knowledge, satisfy the requirements. The next three sections describe the principles behind the formation of the safety specification.

---

[8] The chain is drawn from the perspective of the Machine. Hence, each Machine Hazard may have many causes and many consequences.

3.4 Risk Reduction

The concept of risk reduction defined in IEC 61508 (see Figure 3) for Electrical/ Electronic and Programmable Electronic Systems is applicable to all kinds of accidents. The definition of risk as the combination of consequence (the number of fatalities, serious injuries, etc.) and likelihood of an accident is universal whether due to Machine performance (e.g. failure of a safety critical component) or Machine composition (e.g. hazardous chemical leakage).

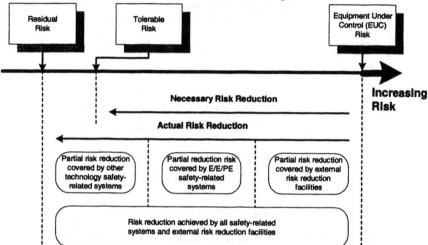

Figure 3: General Concepts of Risk Reduction from IEC 61508

The *necessary risk reduction* is the reduction in risk that has to be achieved to meet the tolerable risk for a specific accident. The safety requirement (what we want to be true in the World) is that the risk of the accident is lower than the unacceptable limit and is reduced as low as reasonably practicable beyond that limit.

The *actual risk reduction* is what is achieved by the design of the Machine or combination of Machines, theoretically that is:

- each Machine meets its specification;
- the combination of Machine specification and domain knowledge satisfies the safety requirements.

If we now relate this concept of risk to the REVEAL model (see Figure 4) we can draw out the safety requirements and specification statements that relate to the Machine at each stage in the risk reduction process. However, measuring the *actual risk reduction* achieved by a Machine or combination of Machines is only possible with statistically significant accumulated Machine operating hours[9]. So we refer to the achieved risk reduction as theoretical.

---

[9] Even then this is only true if the machine is used in precisely the same way as it has been in the accumulated operating hours.

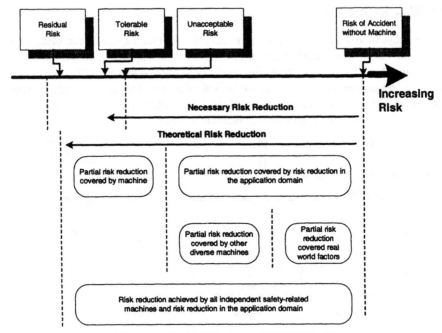

Figure 4: Revised Concept of Risk Reduction

The *necessary risk reduction* can be apportioned to the hazards associated with each Machine by quantifying the hazard model described in section 4. As pointed out in [Fowler 2001], the reduction of risk is not just about failure integrity but also implementation of correct functionality[10]. The achieved risk (R_A) is the product of the likelihood (P_S) that the risk reduction measure works (given perfect integrity) and the likelihood (P_{MS}) that the Machine(s) implementing the measure does not fail. For example, a machine used to detect and mark mines in a minefield needs to be highly accurate in the identification of mines. However, the probability of not detecting mines is limited by the type of mine and the detection technology employed, even if the machine has perfect integrity.

When risk is apportioned to more than one Machine then the overall risk achieved is calculated as:

$$R_A = P_{Si} \times P_{MSi} \times P_{Sj} \times P_{MSj} + P_{CMi\text{-}j}$$

The term $P_{CMi\text{-}j}$ represents the probability of the risk occurring due to a common error, i.e. $P_{CMi\text{-}j} = 0$ if and only if P_{Si}, P_{MSi}, P_{Sj} and P_{MSj} are independent. This need for independence gives us our first safety specification statements. Whilst we may not be able to measure the value of some of the variables, we still have a valid common currency for the setting of safety targets. Nevertheless, we must not ignore terms simply because we cannot measure them.

---

[10] See section 3.7 for more explanation.

3.5 Machine Safety Functions

The REVEAL approach defines a System Specification as a description of the external behaviour of a Machine. Statements in the specification:

- cover only shared phenomena in the interface between the World and the Machine;
- can only constrain shared phenomena that the Machine can control;
- may constrain the design and implementation of the Machine.

Phenomena in the interface are shared where they are in some sense visible to both the World and the Machine. For example, an operator may transmit some voice message to an airborne aircraft. The voice message is shared between the Machine and the World. The Machine creates the voice message, but the World (in this case the aircraft pilot) pays attention to it. It is through this shared phenomenon that the Machine seeks to influence the behaviour of the World. Other shared phenomena will be controlled by the World and responded to by the Machine. Shared phenomena are either controlled by the Machine or controlled by the World.

One objective of the approach is that the specification should not unnecessarily restrict the design and implementation of the Machine. However, with safety there is a legal requirement to use current good practice. There are many restrictions imposed on the Machine's design, development and implementation from regulations, legislation and safety standards. These form both safety requirements and domain knowledge (e.g. good practice). Furthermore, the design and implementation of the Machine can introduce new hazards to the application domain. For example, when the machine uses a human operator, the hazards to that operator need to be considered. For this reason the safety requirements and specification may be revised and added to as the implementation becomes fully known.

The safety specification of the Machine must take into account the method for the reduction of risk and the tolerable likelihood that the realisation of the method will not fail. The Machine Hazards define undesirable behaviour of the Machine with respect to its affect on the World as a result of the Machine's inability to:

- control shared phenomena;
- react to undesirable behaviour of phenomena controlled by the World.

Safety specification statements for the Machine are formed by considering the necessary mitigation for each Machine Hazard. The range of mitigation required is dependent on the type of hazard and the implementation chosen for the Machine. Hazards can be created by either loss or deviation of function or the physical disposition of the Machine.

For the moment we will leave aside hazards associated with physical properties of the Machine (such as sharp edges, hazardous chemicals and weight) and concentrate on Machine functions. It is common practice whilst developing safety-related Machines to concentrate on the integrity level to be assigned to Machine components, rather than derive the safety properties of each function. Although a group of functions may all have the same integrity level, they can have fundamentally different properties. Furthermore, the process of identifying and defining safety functions must also have sufficient integrity.

3.5.1   Protection Functions

Protection functions monitor shared phenomena under the control of the World (e.g. another Machine) and react to defined stimuli by performing some predefined mitigation to prevent a hazard from occurring[11]. These functions are typically expressed in terms of describing what the Machine must *prevent*. The specification may also define what the Machine must do if the primary mitigation fails. This relies on the existence of a safer state to which the application domain can revert whilst retaining degraded functionality or even no functionality. This loss or reduction in functionality may conflict with reliability requirements, so the machine specification needs to include statements on the avoidance of invalid or protracted protection action.

For example, an automatic train control system is monitored by a protection Machine, which requests service braking in the event of the train exceeding its safe movement envelope. If the train continues outside this envelope then the protection Machine requests the emergency brakes bringing the train to a halt. The train can then only proceed in a speed-restricted manual mode. However, this manual mode of operation reduces the performance of the train, which could lead to delays to the passenger service. In this case, the protection Machine should also be specified to minimise both the frequency of invalid protection action and the duration of the protection action.

3.5.2   Control Functions

A control function is one that has to be available in order for the application domain to remain safe, although the function may only be required during certain specific actions or operational modes (e.g. in flight). Typically, this may be to ensure that the system remains in a safe state in response to dynamic changes in the environment, or to provide some functionality that is needed in response to an emergency. Specification statements for control functions are usually described in terms of what the Machine *must* do. Control functions often need to respond to data in real-time in order to maintain safety, and although there may be scope for some graceful degradation of service, there is usually no simple safe state.

Safety-related continuous control functions are more difficult to implement and often require much more complex safety arguments and comprehensive evidence to demonstrate.

---

[11] A machine may be specified only to monitor shared phenomena and raise an alarm, with another machine specified to react to the alarm.

3.6 Machine Safety Specifications

The safety statements in the Machine specification will be derived from a number of sources including:

- the risk reduction process - risk targets and Machine independence;
- Machine hazard mitigation - functional and physical properties;
- constraints on design - standards, legislation, regulation, etc.[12];
- constraints on development or management process - Black box Machine safety issues such as Quality and Safety Management[13];
- the resolution of any requirements or specification conflicts.

The safety statements covering functional and physical properties of the Machine can include:

- each Machine hazard;
- deviation from each control function of the Machine (both desired and emergent) that would give rise to a Machine hazard;
- the tolerable likelihood of occurrence for each control function deviation;
- the protection functions required of the Machine to ensure that phenomena under the control of the World do not give rise to a Machine hazard;
- the tolerable likelihood with which each protection function will not be performed by the Machine;
- the restrictions on physical properties of the Machine that constitute a Machine hazard.

The vocabulary of the function definition must include only terms relevant to the application domain. Functions can be initiated by the World or by the Machine. World initiated events contain two separate descriptions: the application-controlled stimulus and the Machine-controlled result of the function.

Functions can be described using structured English or a mathematical notation. In either case, we are interested in the net effect of the function, not in how it is achieved. Therefore, it is usually a bad idea to specify functions procedurally. Unfortunately, the tradition in safety-related systems design is to specify functions (and hazards) in terms of the Machine without reference to the World (i.e. Machine failure modes). This is not consistent with REVEAL or the definition of hazard and encourages a distorted view of the Machine's affect on the application domain. Consequently, hazards may be overlooked.

---

[12] Design constraints introduced for other reasons may require safety analysis to determine additional safety statements.

[13] In particular, a statement that the machine implementation must be analysed to identify any additional machine hazards. For example, architectural defences that might enhance safety, or negate the need for mitigation elsewhere or features of the machine not envisaged in the original requirements, but that could lead to a hazard.

Whatever notations are used for functional safety specifications, the following elements should be included.

- a function statement:
  - o stimulus - what causes the function to happen;
  - o information used - inputs to the function, or knowledge about the World;
  - o information generated - outputs to the World, or changes in state;
  - o effect - what does it actually do.
- Safety properties:
  - o logical - a precise statement of what should not fail and under what conditions;
  - o performance (including sequence);
  - o tolerance;
  - o assurance requirement for each property, e.g. Safety Integrity Level.

For example, the logical safety properties of the function 'display airspeed' could be defined as 'Loss of', 'too high', 'too low' (perhaps even a definition of what constitutes 'too high' and 'too low'). This level of detail is not always relevant, but is particularly useful for all but the simplest Machine. The amount of evidence needed to prove the property "loss of" is significant compared to "too high" or "too low". It is also possible to miss subtle properties; e.g. the function Emergency Fuel Tank Jettison (EJ) has the property "Loss of ", which could be interpreted as the complete loss of EJ whereas partial loss of EJ would also be hazardous if tanks were fitted on each wing tip, i.e. where it would result in an unsafe shift in the aircraft's centre of gravity.

3.7 Machine Integrity

The international safety standard IEC 61508 [IEC 61508] defines safety integrity as the *"probability of a safety-related system satisfactorily performing the required safety functions under all the stated conditions within a stated period of time"*. However, safety integrity is not just about the integrity of the Machine's implementation but also the integrity of the Machine's specification. In turn, the integrity of the specification is dependent on the completeness and correctness of the safety requirements of the application domain and the integrity of the Satisfaction Argument.

To demonstrate that a Machine is safe it must be shown that:

- all the risks within the application domain have been identified and correctly flowed down to the Machine hazards;
- the combination of safety specification statements and application domain knowledge satisfy the requirements with sufficient integrity[14];

---

[14] For example, the application domain knowledge might include assumptions, which must be proved true.

- the Machine meets its safety specification.

To achieve the latter the Machine must demonstrate that:

- all safety properties have been implemented completely, i.e. the risk reduction or hazard mitigation exists;
- the implementation is effective, i.e. when working perfectly the Machine will mitigate the hazard;
- the likelihood that the mitigation will fail is acceptable.

If the Machine does not implement a mitigation or the likelihood of the mitigation not working (given a perfect Machine) is close to the tolerable limit, then no matter how unlikely the Machine is to fail it will not be safe enough. Frequently the assignment of a Safety Integrity Level is taken only to address Machine failures or the processes with which the Machine is developed. However, the integrity of a Machine is also dependent on the integrity of the Satisfaction Argument, i.e. the completeness and correctness of the safety requirements, domain knowledge and specification. A clear specification with a demonstrated integrity can be used to focus the Machine's development process and aid in the identification of safety-related Machine failures. This is the basis of the White Box Safety Engineering approach developed by Praxis Critical Systems [Simpson 1999].

3.8 The Human Machine

In writing this paper we have aimed to avoid issues to do with the particular implementation options available to a Machine supplier. The requirements and specifications related to a Machine can be developed (at least initially) independent of the chosen implementation. However, it is clearly of concern that not all implementation choices are covered by safety standards, in particular the interpretation of hazard mitigation targets. One example of this is the Machine that relies on human operators.

Many safety critical systems employ human operators who are often used partly or fully as the mitigation of Machine hazards. In the context of REVEAL humans are not Machines in themselves rather human Machines are the combination of the human operator, the procedures, training, job aids and the operating environment. The design of the human Machine needs to take account of all of these elements and the rigour applied to deriving a correct specification, which must include the shared phenomena and implementation of the design must be appropriate to the required Machine integrity. Using humans as part of a Machine also introduces new risks, e.g. the risk to the operator from the rest of the Machine.

The international standard [ISO 13407] defines four basic principles behind a Human-centred design approach together with through lifecycle guidance on their application:

- Active involvement of users and a clear understanding of user and task requirements;
- An appropriate allocation of function between users and technology;

- The iteration of design solutions;
- Multi-disciplinary design.

The ability of the human operator to perform their role is not just about individual behaviour with respect to the human application domain interface, but also the wider framework:

- Team / group - e.g. peer pressure, supervision;
- Organisation and Management - e.g. financial constraints;
- Legal and Regulatory;
- Social and Cultural.

There are many examples of accidents throughout history where safety is undermined by the human Machine. We need to build integrity into the specification of Machines to include ALL aspects of the Machine implementation to manage their interaction and ensure a total system safety approach.

4   Building the Hazardous Event Chain Model

Correctness and completeness of the safety analysis underpins the identification of requirements and the specification of a safety-related Machine. Essential to the identification of a complete set of hazardous event chains in a Machine is the application of a systematic approach. This can be achieved using a combination of top-down deductive (or accident-down) and bottom-up inductive (or implementation-up) analyses.

This approach to accident and hazard identification has been successfully applied by Praxis Critical Systems on a number of projects in various different application domains ranging from Air Traffic Management to Military Aircraft systems. Our experience is that reliance on one or other technique produces an incorrect and incomplete analysis. For example, a huge amount of implementation-up analysis had been carried out in the form of a Functional Hazard Analysis (essentially a list of failure modes) to assess hazards for a passenger aircraft. As part of the Independent Assessment of the project a top-level accident and hazard identification model was constructed, which included input from application domain experts. As a result a number of hazards and issues were identified that had been missed in the FHA.

The following sections present an accident and hazard identification process that when supported by the application of the key principles described, aids the identification of a correct and complete set of hazardous event chains.

4.1 Principles

A number of safety analysis techniques exist and many have been around for a number of years, each with a number of advantages and disadvantages. Our preferred accident and hazard identification process takes the benefits of these techniques and combines them to produce a robust systematic approach.

The approach uses the benefits of HAZOPS combined with the benefits of a systematic top-down approach. The HAZOPS approach uses structured brainstorming sessions with application domain experts that provide the analysis with an implementation-up view. When combined with the rigorous nature of a systematic top-down approach this provides greater confidence in achieving a correct and complete set of hazards.

4.1.1 Inductive Analysis

The word induction is defined here as 'reasoning from particular to the general'. In the context of a Machine and by a process of induction we can ascertain from the analysis of the shared phenomena the effect of functional deviation on system operation. Many approaches to inductive system analysis have been developed. Some examples of inductive safety analysis techniques are: Hazard and Operability Studies (HAZOPS), Failure Mode and Effect Analysis (FMEA), Failure Mode Effect and Criticality Analysis (FMECA), Fault Hazard Analysis (FHA), and Event Tree Analysis (ETA).

4.1.2 Deductive Analysis

The word deduction is defined here as 'reasoning from the general to the particular'. In the context of the Application Domain, considering a general undesired event i.e. an accident, particular initiating events can be ascertained using a process of refinement through deduction. The distinction in approach with an inductive method is to analyse "what needs to happen" rather than "what can fail". Accident investigations would use a deductive approach based on the observed undesired event (i.e. the accident). The investigators would ask: "What chain of events caused the aircraft to hit the ground?"

An example of a deductive analysis technique is Fault Tree Analysis (FTA) as defined in [NUREG 0492][15].

4.1.3 The "Immediate Cause" Principle

Essential to the deductive process in identifying the causes of an event is the identification of the **immediate and necessary and sufficient** causes for the occurrence of that event. Completing the next level by jumping directly to the **basic** causes must be avoided even if these may appear obvious.

- **Immediate** cause is the next level of detail without intervening context or substance. It does not mean occurring, acting, or accomplishing without loss or interval of time.
- **Necessary** causes are the logically unavoidable events required to occur, whether in a particular sequence or a defined time frame, for the event to occur.
- **Sufficient** cause implies consideration of only those events needed and stated in such a manner as to impart a precise description of the event.

---

[15] FTAs can be constructed inductively, but this reduces the value of the analysis.

For example, a system to control a safety critical release output is implemented using a programmable electronic device. One of the undesired top events is "extraneous release of Equipment Under Control (EUC)" (i.e. release of the EUC when not required or intended). The cause of this event could be written as "extraneous output enable", which is logically correct but not complete. The **immediate and necessary and sufficient** cause for this event should be written as "extraneous EUC release pulse (X volts for > Y seconds) ANDed with the any necessary conditions such as; EUC fitted or System Operational[16].

Without precision in the definition the analyst may be tempted to define the next level of cause as "Output from programmable electronic device at incorrect time", but this may lead to other basic primary and secondary causes being overlooked, e.g. EMC-induced voltages or connector/circuit board short circuits, etc.

4.2 What We Do

The systematic but restricted nature of inductive analysis tends to produce comprehensive but incomplete definitions of the Machine hazards. This is due to the amount of information to be analysed and a focus on Machine failures rather than the shared phenomena. For example, for a component of a system a selection of failure modes (or guidewords) is postulated. However, the list of failure modes is simplified and does not consider all credible failure modes, or combination of, otherwise this would create an excessive amount of work. Most of the end effects would have little or no bearing on safety and this makes the process inefficient. Furthermore, the magnitude of the task makes it tedious and analyst lapses are common, for example cut and paste errors. Summaries are hard to produce without loss of subtle differences that are important.

The creative nature of a deductive analysis tends to produce a whole variety of results dependent on the knowledge and skill of the analyst. However, careful adherence to the method tends to produce a far more accurate analysis with a greater degree of completeness that is easier to review. By focussing on the safety issues we have found that unnecessary analysis of other effects is minimised.

Both approaches are necessary in order to ensure completeness and correctness and therefore there must be traceability between the various analyses. There is often no direct link between the results of inductive and deductive analysis and the following should be taken into account when doing so.

- It is common for deductive analysis to postulate properties of shared phenomena not considered by inductive analysis, for example due to the use of simplified failure modes.
- Deductive analysis must incorporate an inductive perspective to ensure functional and architectural completeness and assess hazards introduced by the Machine's implementation or functionality.

---

[16] The conditions are often assumptions (i.e. domain knowledge) about the World, which need to be proved true.

When using the results of inductive analyses like HAZOPS it is important to ensure that the immediate cause principle is not usurped by the desire to just add in the additional events. As well as single event scenarios, inductive analysis can also identify potential multiple event chains. In either case, the deductive analysis should be re-assessed to establish where and why the events were missed. The resultant modification may reveal further scenarios not previously considered.

The analyses can and should be supported by the use of distinct perspectives or models of the Machine and its context. One or more perspectives can be constructed to assist the analyst in identifying further scenarios. For example:

- Functional - the functions and functional relationships;
- Architectural - the electrical, physical and logical realisation of the system and respective inter-relationships;
- Human Interaction - the interfaces either directly or indirectly between the system and human operators and maintainers;
- Data - the types of data manipulated or managed by the system, this might include modelling the timing or sequencing of data flows;
- Networks - the communication channels (virtual and physical) that link the system functions, data flows and architecture elements.

By taking this approach and using the appropriate tool for modelling it is possible to achieve completeness in the breadth of the analysis of a system.

4.3 Where Do We Start?

The top-down accident and hazard analysis approach uses refinement to assist in the identification of hazardous event chains. The process requires input from application domain experts and begins with the identification of all top-level accidents of concern within the Application Domain.

For example, the accident mid-air collision can be refined to loss of separation in the horizontal and loss of separation in the vertical, in a manner comparable with requirements refinement, as shown in Table 1.

$A_{MID-AIR}$ $\downarrow$	Aircraft hits other aircraft in flight	R $\downarrow$	Aircraft shall be kept a safe distance from other aircraft
H_{VS}	Loss of safe vertical separation between aircraft in flight	R_{VS}	Aircraft shall be separated by more than the vertical separation minima in the vertical plane.
OR		AND	
H_{HS}	Loss of safe horizontal separation between aircraft in flight	R_{HS}	Aircraft shall be separated by more than the horizontal separation minima in the horizontal plane.

Table 1: Example Safety Requirements Refinement

We have found Fault Tree Analysis (FTA) as being the most appropriate method for assessing the hazardous event chains leading from accidents. The FTA technique was developed around the immediate cause principle.

The analysis method uses a systematic deductive approach in the construction of the fault tree, focusing on "what needs to happen" rather than just "what can fail", in a step by step refinement of the top-level accidents. As the interface and thus the shared phenomena are identified so further refinement can take place down to the Machine hazards.

The top undesired event is *Death or Injury within the application domain*. At first this may seem at a very high level but the tree quickly develops down to events at the boundary of the Machine. Figure 5 shows the refinement of this event to the next level using the immediate cause principle.

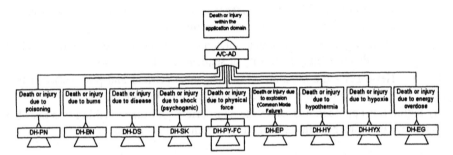

Figure 5 - Top level Accident Fault Tree

Whilst it is not always strictly necessary to start at this high level the choice and precise definition of the top event is critical to the success of the analysis. An incorrect or incomplete top event will in most cases, invalidate the whole analysis. The greater the vagueness or the lower the level[17] of the top event description the greater the likelihood of incompleteness in the analysis. For this reason, the top event definition must be carefully considered and accurately and unambiguously stated before modelling is started. The top event description may need to be re-assessed as the analysis progresses or more detail becomes known, to ensure it remains correct.

To ensure the validity of the top events the identification process should include expert knowledge from the application domain.

The event Death or injury due to physical force is often the major branch of the tree as this will include accidents such as collisions and derailments. Figure 6 and Figure 7 show the development of this branch of the FTA for an aircraft using the principles and approach discussed in this paper. Note that for brevity only part of the tree is shown.

---

[17] There is a tendency to start with system component or functional failure modes rather than establish the undesired property of the system. Starting at this lower level could mean missing important system function properties relevant to hazards.

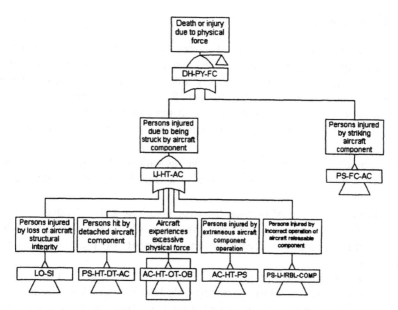

Figure 6 - FTA showing development of the Physical Force event (1)

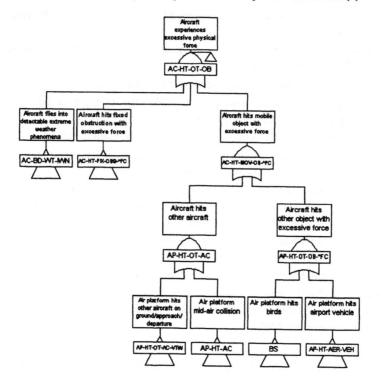

Figure 7 - FTA showing development of the Physical Force event (2)

5 Conclusions

Our stated aim is the development of a more rigorous method for the identification of safety requirements for safety related systems providing greater assurance that these systems are specified completely and correctly. A significant amount of experience has been gained from our Requirements Engineering approach REVEAL, which provides clear benefits to our aim not just in terms of a better assurance of safety but also in managing the safety requirements of complex systems.

The approach does require more effort, but as with our Requirements Engineering experience this effort and rigour can play a significant part in reducing overall project risks.

We address the nature of systems by using notations and techniques that are not dependent on whether subsystems are software, mechanical, process, etc. We have applied this approach successfully on a number of projects including the identification of:

- functional safety properties for Commercial Off The Shelf Real Time Operating Systems for Integrated Modular Avionics based architectures;
- safety indicators for assuring the safety of Reduced Vertical Separation Minima airspace;
- safety requirements for components of an Airborne Early Warning System.

We have briefly described just some of the aspects of REVEAL that are relevant to our stated aim. These help us to:

- distinguish between requirement and specification;
- capture domain knowledge to make the connection between what a system does and the requirements;
- define successful safety arguments by demonstrating that:
  - o we have a complete and correct set of safety requirements (with the risk budget balanced across ALL risks);
  - o the external risk reduction mechanism used to determine the proportion of the risk budgeted to the Machine (i.e. part of the domain knowledge) is true[18];
  - o the combination of specification statements and domain knowledge satisfy the safety requirements;
  - o the Machine satisfies its safety specification.

We need to build integrity into the specification of Machines to include ALL aspects of the Machine implementation to manage their interaction and ensure a total system safety approach.

---

[18] For a risk that is reduced by a number of machines then those machines must be independent, i.e. the risk of any potential common error must be reduced by the necessary risk not the apportioned risk.

6 Acknowledgements

This paper is based on the approach to identifying safety requirements developed by Praxis Critical Systems on a number of large-scale programmes. The approach was also adapted as part of the research study into the certification of COTS Real-Time Operating Systems for Advanced Avionics Architectures, conducted by Praxis Critical Systems on behalf of the Defence Evaluation and Research Agency.

The views expressed in this paper are those of the authors and do not necessarily represent the views of any the organisations mentioned throughout. The authors are, however, grateful to those people in these organisations who have helped form, revise and improve various aspects of the approach.

In particular, we would like to thank Dr. Andrew Vickers, Rosamund Rawlings, Vicky Brennan, Martin Croxford, Dr Trevor Cockram and Efi Raili for their contribution.

7 References

[Def Stan 00-56]   Ministry of Defence: Safety Management Requirements for Defence Systems, December 1996

[Fowler 2001]   Fowler, D, Tiemeyer, B, Eaton, A: Safety Assurance of Air Traffic Management and Similarly Complex Systems, Proceedings of the 19th International System Safety Conference, September 2001.

[HSE 1995]   Health and Safety Executive: Out of Control; HSE Books ISBN 0717608476, 1995.

[HSE 1999]   Health and Safety Executive: Reducing Risks, Protecting People, HSE Discussion Document DDE-11, 1999.

[IEC 61508]   subcommittee 65A: System aspects of IEC technical committee 65: Industrial-process measurement and control: Functional safety of electrical/electronic/programmable electronic safety-related systems, December 1998.

[ISO 13407]   ISO Technical Committee ISO/TC 159 "Ergonomics", Human-centred design processes for interactive systems, June 1999.

[Jackson 1995]   Jackson, M: The World and the Machine, Proceedings of the 17th International Conference on Software Engineering, IEEE, pp.283-292, 1995.

[Leveson 2001]   Leveson N: Evaluating Accident Models using Recent Aerospace Accidents, Software Engineering Research Laboratory MIT, 20 June 2001.

[Lutz 1993]   Lutz R: Analyzing Software Requirements Errors in Safety-Critical Embedded Systems, IEEE international symposium on requirements engineering, San Diego, IEEE Comp Soc Press, 1993.

[NUREG 0492]   US Nuclear Regulatory Commission: Fault Tree Handbook, January 1981.

[REVEAL]   Praxis Critical Systems: REVEAL A keystone in Modern Systems Engineering, available from Praxis Critical Systems email: reveal@praxis-cs.co.uk.

[Simpson 1999]   Simpson A, Ainsworth M: White Box Safety, Proceedings of 13th ERA International Avionics Conference, 1999.

[Vickers 1996]   Vickers, A J, Smith, J, Tongue, P: Complexity in Requirements Engineering, Proceedings UK INCOSE Symposium, 1996.

[Yellow Book]   Railtrack: Engineering Safety Management, Issue 3, Yellow Book 3. ISBN 0 9537595 0 4, 2000.

RISK

Integrated Design Analysis

Jack Crawford
Crawford Consultancy Limited
3 Green Lane, Farnham, Surrey, GU9 8PT, UK
Tel: +44 1252 714983
Email: jpk@crawfd.co.uk

Abstract

A methodology for improving the consistency and completeness of hazard analyses, and hence the integrity of predictions based on them, is described.

1 Introduction

This paper extends the ground covered by a paper presented at the Ninth Safety-critical Systems Symposium on "Ways of Improving our Methods of Qualitative Safety Analysis" [Crawford 2001]. It will be argued that we need to do much more than is customarily done to co-ordinate the various forms of analysis which we use, such as Fault Tree Analysis (FTA), Hazard and Operability (HAZOP) Studies, Failure Mode and Effects (and Criticality) Analysis (FME(C)A) and others. If the analyses are inconsistent with one another, it follows that they are also incomplete.

It will also be argued that our ability to predict probabilities of accidents has considerable room for improvement, because current methodologies largely ignore the question of stability in the behaviour of systems. Instabilities can greatly increase the probability of accidents.

Integrated Design Analysis (IDA) is a set of concepts designed to promote improvements in those areas. A description of IDA was first published in Ordnance Board Pillar Proceeding P120(1) [MOD 1996/1], referred to hereafter as *Proc P120*. It applies the concepts in the limited but demanding field of safety, arming and fuzing systems for munitions. The concepts are here extended to a wider range of analytical methods, and the scope to a wider range of systems.

2 Why Integrate the Analyses?

Reasons for integrating the various elements of a safety analysis have been given as follows [Marshall 2000]:

- The increasing complexity of both designs and design standards increases the probability that a safety critical aspect of a design will be overlooked.

- Consequently there is a need to improve on current procedure, especially in terms of discipline.

- It makes sense to improve the way we use existing analytical tools, such as FTA and FMEA, by relating them to each other. This should ensure that we are able to create, understand and communicate a complete picture.

3 Developing the Analyses from a Common Root

As an aid to achieving consistency, IDA requires that all branches of the analysis of a system should be developed from a common root. This applies to analysis of the correct functioning of the system as well as to the analysis of causes, modes and consequences of failures. The essential characteristic of the common root is that it should convey a complete understanding of the way the system is designed to work. In *Proc P120* it takes the form of a "Basic List of Items and Functions" (BLIF). The "Items" are a list of the components of the system. Depending on its significance for safety, an item may be a sub-system, an assembly or even a single component. The worked example in *Proc P120*, of a munition fuzing system, breaks the system down to a low level since many of the components are safety-critical. Its first item is shown in Figure 1.

ITEM	FUNCTIONS
1. FUZE BODY (Includes Magazine Housing).	a. Provides tension link between PLUNGER (2) and bomb case. b. Protects internal components. c. Locates internal components. d. Provides rf screen for internal components. e. In the safe state, combines with the SHUTTER (14) and MAGAZINE (20) to protect the downstream elements of the explosive train from the effects of a premature function of the DETONATOR (19) and/or PYROTECHNIC DELAY (18). f. Secures the fuze to the bomb case.

Figure 1. An item in a Basic List of Items and Functions (BLIF).

The format is convenient in that it is ready to be used as the first two columns of an FMEA worksheet. The main point to note is that the functions of the Fuze Body are described more fully than is usually done in an FMEA. They include functions performed in combination with other components, the latter being identified by their name and serial number. The result makes it clear that the Fuze Body, although it is only a passive component, has as many as six functions.

Experience of using IDA has shown that it needs considerable thought and care to identify all the functions of components, and especially of mechanical and structural components. The functions of electronic components tend to be fewer and simpler. The reason for taking so much trouble over this stage of the analysis is that, if we are unaware of some function of a component, we shall be unlikely to see how that function might be affected by failures within the system. The consequences of missing such a trick can include unnecessary costs and may even be hazardous. It is cheaper to introduce an improvement at the design stage than at later stages of development, and far cheaper than paying for modifications, and perhaps an accident, during service.

4 The Use of Frames

One of the problems in assessing safety is that, when we have done our best, we still have no way of knowing whether we have thought of everything. The psychological concept of the "frame" has been found helpful as an aid to improving the completeness of safety assessments. Bateson describes it in terms of two analogies: "the physical analogy of the picture frame and the more abstract ... analogy of the mathematical set" [Bateson 1987]. In applying the concept to the practical task of assessing the safety of explosive devices, *Proc P120* describes a frame as "a point of view from which to look at a system" and also as representing "the area of validity of a model" (having defined a model as "any representation of the system which is to be assessed").

A significant reason why potential causes of accidents have been overlooked in safety assessments is that there was some point of view from which the system was not considered. The aim of a "frame analysis" is to reduce the likelihood of that shortcoming by setting out in advance all the points of view from which the system should be examined in order to achieve a complete assessment. In the case of safety, arming and fuzing systems, *Proc P120* identifies a set of 9 frames:

- Chemical interaction.
- Circuitry.
- Computing systems.
- Construction.
- Environmental protection.
- Operating logic.
- Power supplies.
- Qualification of explosives and explosive components.
- Sterilisation.

That list is an example of a generic set of frames. While it will be suitable for many devices of the class, a generic set of frames needs to be reviewed, and modified if necessary, before being used in any particular case. For example, some fuzes have no computing systems or power supplies. Others might be designed so that their changes of state may be controlled by telecommunication, which would require the addition of a "Communications" frame.

In IDA, each function of each component of the system is related to one or more frames. Figure 2 reproduces another item from *Proc P120*. This item has five functions, of which three are in the operating logic (OL) frame and two in the circuitry (CR) frame. Where conformance to a design code is required, its clauses may be similarly related to frames as an aid to checking whether the functions of the system components satisfy the requirements.

ITEM (a)	FUNCTIONS (b)	FRAME (c)
13. DELAY ACTUATOR (Includes Locking Pin, Cable and Connector).	a. In the safe state, locks the SHUTTER (14) in the out of line position even if arming force is exerted by the ARMING LEVER (11).	OL
	b. On receipt of bomb release signal, initiates 5 s delay.	OL
	c. At end of the arming delay, withdraws the Locking Pin (13).	OL
	d. Cable transmits arming signal from aircraft.	CR
	e. Connector provides interface between Cable and aircraft system.	CR

Figure 2. Example of a Frame Analysis.

The concept of frames was inspired by the way we naturally think about things, i.e. from various points of view. The purpose of using frames in this more formal way is to help us not to miss tricks.

5 Integrating FTA and FMEA

It has been argued in *Proc P120* and elsewhere [Crawford 2001] that there is a simple logical relationship between FTA and FMEA which has been overlooked in the safety and reliability literature and standards. The relationship may be deduced from the theorem that when we analyse any clearly defined system, we are analysing a single set of potential failure modes. Given that the FMEA encompasses the whole system, it follows that the failure modes shown in any fault tree for the system must be a sub-set of the total and should all appear in the FMEA. Conversely, all failure modes identified in the FMEA, which could potentially contribute to one or more fault tree top events, should be included in the FTA.

That proposition has by now been aired fairly widely, and so far nobody has argued against it. But how much does it matter? An example of what can happen when the relationship is not recognised was provided by a major supplier of propulsion systems. The safety assessment report for one of its products included a FMECA and an FTA of 11 fault trees. A consistency check showed that the FTA contained 46 potential failure events which were not recorded in the FMECA, after eliminating multiple counting of events which appeared more than once. The FMECA contained 23 potential failure events which were not shown in the FTA, but which could have contributed to one or more of the top events. It is evident that each analysis, by being so inconsistent with the other, was also far from complete. Readers may find it interesting to try this exercise on a sample of the analyses available to them.

That example also suggests that when we think *inductively* (as when working on a bottom-up analysis such as FMEA) we may to some extent think of different ideas to those which come to us when we think *deductively* (as when working on a top-down analysis such as FTA). Whatever the explanation, the evidence indicates that we have an untapped cognitive resource which can be exploited to help us to improve the completeness of our analyses.

A method of co-ordinating the two forms of analysis is described, with a worked example, in *Proc P120*. A more compact example, good enough for the purpose of this paper, is drawn from classical mythology. Readers may recall Theseus' expedition to slay the Minotaur, the monster who was kept in a labyrinth, living on human sacrifices. Nobody who had entered the labyrinth had ever found the way out. Fortunately the handsome Theseus attracted the fancy of the king's daughter Ariadne who provided him with a means of making his escape after killing the monster, who would only be vulnerable if caught asleep in the small hours. The escape kit consisted of a reel of silken cord, to be unwound from the labyrinth entrance so that Theseus might retrace his steps later.

If Theseus had had to conform to modern regulations for safety-critical devices, he might have performed a quick FMEA and FTA on the lines of Figures 3 and 4. His IDA skills would have led him to ensure that each failure mode identified in the FMEA was allocated either to a fault tree or to a "Set R" of residual events (the latter often being reliability-critical). In this case Figure 3 identifies two potentially fatal failure modes and a non-hazardous one leading to mission failure. Conversely he would have ensured that each failure mode identified in the FTA was included in the FMEA, and traceable by some form of cross-reference as shown in Figure 4. By that method, consistency between the analyses is ensured and completeness is improved (at least in more complex examples).

172

ITEM	FUNCTIONS	FAILURE MODES	FAILURE EFFECTS	ALLOC-ACTION
1. Silken Cord.	To guide Theseus out of the labyrinth.	Cord too short. End reached before Minotaur is found.	Theseus forced to withdraw. Mission failure.	R
2. Reel.	To facilitate laying of the cord.	a. Spindle squeaks when reel is rotated.	a. Minotaur, awoken by noise, overwhelms Theseus.	FTA
		b. Spindle breaks. Reel falls and clatters on floor.	b. Minotaur, awoken by noise, overwhelms Theseus.	FTA

Figure 3: FMEA of Escape System, cross-referenced to the FTA.

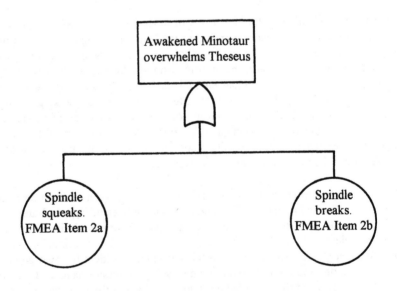

Figure 4: FTA of Escape System, cross-referenced to the FMEA.

So far, of course, only the first four columns of a normal FME(C)A worksheet have been used, because they are all that is needed for co-ordination with other branches of the analysis. When that has been done, the rest of the FME(C)A is completed in the normal way.

6 Integrating a Manufacturing Process FMEA

Proc P120 recommends that a manufacturing process FMEA should be conducted separately from the product design FMEA, because there has been a lack of rigour in analyses submitted to the MOD which have tried to combine the two. But it does not explain how the former should be integrated.

Before going further, it may be useful to be clear about the difference between those two kinds of analysis. In principle, the difference is:

- A design FMEA assumes that the product will be correctly built according to the design, but questions whether the design will satisfy the requirement.

- A process FMEA assumes that the design will satisfy the requirement, but questions whether the product will be correctly built.

In practice, the distinction may not be quite as sharp as the principles suggest. For example, designers have to recognise the limitations of the manufacturing system and refrain from demanding capabilities which the processes cannot offer.

The method of integrating a process FMEA is similar to that for a design FMEA. It is illustrated by the extracts from each type of analysis at Figures 5 and 6, which are based on examples given in a motor industry standard [CFGM 1995]. The list of Processes and Requirements from which the process FMEA is developed is analogous to the BLIF in a design FMEA. The allocation of failure modes and effects either to a fault tree or to a set of residual events is identical, as is the need for cross-references between the FTA and the FMEA. This procedure allows the FTA to include both product and process failure modes, as is necessary when quantification is required, while keeping the two kinds of FMEA separate.

The motor industry standard requires that both kinds of FMEA be done by teams, and that design teams include process representatives and vice versa. That policy should satisfy the need to ensure that the effects of process failure modes on the functions of system components are recorded, provided that the design FMEA is available when the process FMEA is being done.

7 Integrating a HAZOP Study

The original concept of a HAZOP study was directed as much to improving the efficiency of industrial processes as to identifying their hazards [Kletz 1992]. In Defence Standard 00-58, its purpose is limited to identifying "potentially hazardous variations from the design intent" and its scope is limited to "systems containing

Item	Functions	Failure Mode	Failure Effect	Alloc-ation
Front Door R.H.	To protect occupant from weather, noise and side impact	Hinge pins shear on impact	Door detaches on impact	FT 1
	To provide proper surface for paint	Corroded exterior door panel	Poor adhesion of paint	R
	Support anchorage for door hardware including mirror, hinges, latch and window regulator.	Corroded interior door panel	Impaired function of interior door hardware	R

Figure 5. Fragment of a Design FMEA showing allocation of failure modes to sets.

Process	Requirement	Failure Mode	Failure Effect	Alloc-ation
Weld hinges to Front Door R.H.	To achieve impact-resistant attachment of hinges to door frame	Weld metal weak due to voids	Door detaches on impact	FT 1
Clean and zinc coat exterior door panel	To provide durable surface for paint	Incomplete coverage of surface	Poor adhesion of paint	R
Apply wax inside Front Door R.H.	To cover inner door surfaces at minimum wax thickness to retard corrosion	Insufficient wax coverage	Interior corrodes - impaired function of interior hardware	R

Figure 6. Fragment of a Process FMEA, corresponding to Figure 5.

programmable electronics" (so that "flow" means "data flow") [MOD 2000]. This paper takes a leaf out of each book by accepting the limitation to identifying hazards, but allowing free rein to the scope.

Proc P120 does not cover the integration of HAZOP studies. The following offers a provisional set of guidelines to be reviewed in the light of experience. The HAZOP procedure is inductive and in that way similar in principle to FMEA. The process of integrating it with other analyses is also similar but, since it focuses on flows and deviations rather than component failures, there are differences in detail. The relationships to other analyses may be summarised as follows:

- Where a HAZOP study is the means of carrying out a Preliminary Hazard Analysis (PHA), as required by Defence Standard 00-56 [MOD 1996/2], it will identify top events for the FTA.

- A cross-check should be made to ensure that causes identified in the HAZOP study are included in the relevant parts of the FTA. Conversely if the FTA is done independently in the first place, as it should be, a check should be made that deviations in material or data flow identified in the tree(s) have been recognised in the HAZOP study.

- If an FMEA is done as well as the FTA, there may be no need to co-ordinate it directly with the HAZOP study. If the FTA is consistent with the HAZOP, the standard process of FMEA/FTA co-ordination described earlier will be enough.

- Within the scope of the HAZOP study, the logic models described in Defence Standard 00-58 may serve instead of the operating logic tree (see below) generated from the BLIF, but this is a matter for judgement in each case.

8 Integrating Logic Analysis

The form of logic analysis used in *Proc P120* is the operating logic tree (OLT). It is structured like a fault tree and so has the advantage of using a familiar notation. It differs from a fault tree in that:

- The base events, instead of being fault events, are the inputs which the system is intended to receive.

- From the base events, the tree models the logic of the system.

- The top event is a desired state or output rather than a disaster.

The OLT is used:

- To check that the system design conforms to the OL requirements of the relevant safety standards.

- To identify states such as ready, safe, alerted, armed, inhibited.

- To identify safety features and check that they correctly inhibit or permit functions and changes of state.

- To track the sequence of operations from inputs to outputs.
- As an aid to studying the degrees of dependence or independence of active components and their functions.

To ensure its integration with other elements of the analysis, the OLT is generated from the BLIF and frame analysis. Figure 7, an extension of Figure 2, illustrates the method used in *Proc P120*. The frame analysis identifies each function which has a part in the OL. Each function, other than the initial state, is given a number and the input required to initiate it is recorded. The data shown in Columns (d) and (e) of Figure 7 are sufficient to enable a tree to be drawn. The fragment of the system OLT which is generated from the OL functions of Item 13, the Delay Actuator, is shown in Figure 8.

The symbol used in Figure 8 for the "Delay Actuator initiated" and "Locking Pin withdrawn" events signifies a command event. IDA also uses it to signify command events in fault trees, to facilitate comparison with the OLT.

A feature of the OLT is its power as a means of communicating the way a complex system works [Marshall 2000]. Beyond its originally intended uses, a further use has been found. It is sometimes the case that designers, when incorporating a commercial off-the-shelf (COTS) component into their design, have only an imperfect understanding of how the COTS item works. The discipline of having to draw its OLT provides a rigorous test of understanding and has been found to be enlightening [Marshall 1999].

9 The Stability Factor

Powerful methods of qualitative and quantitative analysis have been developed during the last half century. In many areas, such as civil aviation, they have supported substantial improvements in reliability and safety. In other areas, such as software, they have helped us to keep abreast of ever-increasing complexity. Yet we continue to be surprised and disappointed and, especially when we are eye-witnesses, horrified by accidents large and small. With the wisdom of hindsight, many of them are seen as "accidents waiting to happen".

To assist our analytical tools to realise their full benefits, we need to pay more attention than we have been doing to the question of stability of system behaviour. While the concept of stability is well developed in the mechanical sense, and by no means new in the statistical sense, it has been under-used in safety assurance and assessment. Yet many would probably agree, when they come to think about it, that there is little hope of predicting the level of safety of an unstable system.

Keynes recognised that point many years ago. Noting how heavily our predictions rely on inductive arguments based on historical records, he wrote: "No one supposes that a good induction can be arrived at merely by counting cases. The business of strengthening the argument chiefly consists in determining whether the alleged association is *stable* ...." (his emphasis) [Keynes 1921].

ITEM (a)	FUNCTIONS (b)	FRAME (c)	NO OF OL FUNCTION (d)	INPUT(S) REQUIRED (e)
13. DELAY ACTUATOR (Includes Locking Pin, Cable and Connector).	a. In the safe state, locks the SHUTTER (14) in the out of line position even if arming force is exerted by the ARMING LEVER (11).	OL		Initial state.
	b. On receipt of bomb release signal, initiates 5 s delay.	OL	13/1	Bomb release signal.
	c. At end of the arming delay, withdraws the Locking Pin (13).	OL	13/2	13/1
	d. Cable transmits arming signal from aircraft.	CR		
	e. Connector provides interface between Cable and aircraft system.	CR		

Figure 7. Example of data required for generating the OLT.

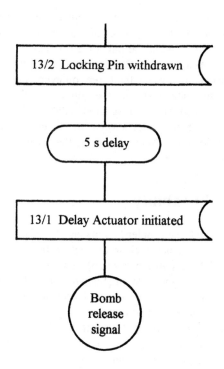

Figure 8. Fragment of system OLT, generated from Figure 7.

There are many potential causes of instability which could invalidate our predictions of the likelihood or frequency of accidents. They include:

- Design errors not recognised in the original safety assessment.

- Failure modes not anticipated in the original safety assessment.

- Assumptions, valid when the assessment was made, being invalidated by later but unnoticed changes in circumstances.

- The responses of people to their perceptions of risk. For example, a lapse into complacency after an accident-free period.

- Errors of management, especially those which, however unintentionally, have the effect of putting people under pressure to take short cuts or to conceal information.

- A weak safety culture.

- Unexpected interactions between the system and its environment.

Readers may well be able to add to this list from their own experience.

The most obvious conclusion to be drawn from those factors is the reality of the unexpected. "Unexpected" however does not always mean the same as "could not have been expected", so it should be part of a prudent safety policy to expect the unexpected. Palmer describes how Bill Medland, a member of the DERA team at Boscombe Down, argued that every fault tree top event should have under it an OR gate. One input to the gate would be the conventionally deduced fault tree, the other being an event called "The Unexpected" [Palmer 2001]. Those who would quantify the tree had then to assign a value to the latter event. Since that value amounted to a claim for their ability to predict the future, they would have to put their personal credibility on the line.

10 Managing Instability

How then should the stability problem be managed? Shewhart, a contemporary of Keynes, was the first to make a clear distinction between "common causes" of variation (i.e. those inherent in any system) and "assignable causes" of variation which are not part of a constant system [Shewhart 1931]. The former account, for example, for the inevitable piece-to-piece variations in a run of manufactured parts. They characterise the stable operation of a system within constant limits of variation, a state he defined as being "in control". The latter are outside influences which may from time to time upset the stable operation of the system; Shewhart defined their characteristics as being that:

- They are not predictable from the record of past behaviour, but

- They may be found and eliminated.

To support the latter assertion, Shewhart devised a set of statistical tools ("control charts") to assist in distinguishing between the two types of cause. He also demonstrated that a state of "control" or predictability does not come naturally to a man-made system. Rather it is an achievement, gained by painstaking work to find and eliminate the assignable causes of instability. It follows that continuing vigilance is needed to maintain, and from time to time restore, stability in the face of threats some of which have been listed above.

Shewhart's insights were extended by Deming who showed that they applied not just to manufacturing but to services and all other forms of work, and especially to management [Deming 1993]. In the context of safety, Cutler has applied the concept of stability to Probabilistic Risk Analysis (PRA) [Cutler 1997] and to accident prevention more generally [Cutler 1999]. Among other points, he argues that differences attributable to the choice of one statistical method or another are swamped by the much greater effects of instability. His readers must surely feel persuaded that stability is a necessary condition for PRA to be credible. It is not so clear that it is sufficient to support the precision of numerical probabilities, especially in the light of questions that have been asked about the credibility of PRA [Crawford 1999, 2000]. Whatever the outcome of that debate, Cutler's initiative has been valuable in highlighting a significant contributor to accidents which has been ignored in the safety and reliability literature and standards.

An example of an assignable cause which greatly increased the probability of an accident was the "disease of sloppiness" which infected the owners of the *Herald of Free Enterprise* and spread down through the company. Measures which could have maintained a low probability of a ferry leaving port with its bow doors open had been derided by top management. We have seen probabilities of failure substantially raised by similar causes in nuclear power, nuclear material processing, the railways, airport security, offshore oil and gas, oil refineries, chemical plants, structural engineering, surgery, radiology, rocketry and other fields. How should so insidious a risk be assessed? It might be considered subversive to suggest that some of our fault trees ought to include a failure mode such as "Complacency in the Boardroom", however harshly realistic it could be. Nevertheless, ways of helping people to be alert to that and other causes of instability need to be introduced to our methods.

Cutler proposes the use of Shewhart's control charts as tools for monitoring stability. For those unfamiliar with their use, Wheeler provides a clear and concise account [Wheeler 1993]. It would not of course be helpful to monitor behaviour at the level of the frequency of accidents; that would be an "after the event" mode and data would only become available (we hope) infrequently. The charts should be used to monitor the frequency of lower level contributing events, and should include near misses. The selection of key events to be monitored could be made with the aid of an FTA, supported if necessary by the use of a suitable importance measure. Even a sceptic should be able to accept the latter as a valid use of quantified methods, given its limited purpose.

The methodology of a truly integrated analysis should recognise stability as a factor, since we cannot expect to improve the integrity of our safety predictions without taking it on board. Safety assessments should indicate critical aspects of system behaviour which need to be monitored during service.

11 Use of Judgement in Applying IDA

IDA involves a considerable amount of work in return for its benefits, and there is still a lack of software that is suitable for taking the donkey-work out of it. In many cases it will not be necessary or practical to apply it to all parts of the system or at all levels. The most logical approach is to apply it in the first place at a high level. The results will allow judgements to be made on where it should be focused at more detailed levels.

Alternatively the output of another high level analysis, such as a PHA or a HAZOP study, may be found usable as an indicator of areas where IDA should be applied. In making such judgements it is worth remembering that the *principles* of IDA, summarised at the end of this paper, are applicable to all parts of the system.

Further judgements are required to decide the selection of methods to be used in any particular case. Readers will be able to improve on the following over-simplified summary of the strengths and weaknesses of the analytical methods considered in this paper:

- The BLIF is an auditable statement of the analyst's understanding of what the elements of the system are intended to do. Its purpose is to provide the foundation upon which further analytical structures may be built.

- FMEA encompasses all parts of a system, including passive as well as active components. As an analysis of consequences, it is an inductive (bottom-up) process. It does not readily show up the effects of combinations of failure modes. It is often done by an individual, although it should be done by a team.

- FTA is strong on displaying the combinations of failure modes which could cause the top event(s). As an analysis of causes, it is a deductive (top-down) process. It is limited to those components of the system which are potential contributors to the top event, other components being ignored. Even a set of fault trees will seldom encompass all components of the system.

- A HAZOP study is an analysis of flows of material or data, and of deviations from design intent. It therefore focuses on active components of the system. It is an inductive process. It is always a team effort. It may miss the effects of failures of passive components.

- The OLT clarifies the behaviour of the system as the design intends, again focusing on the active components. It can be used to highlight dependencies between active components, but is unlikely to show those involving passive ones. Although it uses the FTA notation, the process of constructing an OLT, as described in *Proc P120*, is inductive.

- The set of frames is an aid to clear thinking about a system. It guides us on the points of view from which the system should be considered. It also helps to remind us of the limitations of the models used in the course of an assessment.

- Control charts monitor selected aspects of system behaviour over time. Their purpose is to indicate signs of instability, in time for action to be taken to find and eliminate the cause. Wheeler characterises the control chart as an aid to detecting signals amongst noise [Wheeler 1993].

The above is only a short list of analytical methods, most of them being in common use. Other methods may be brought in and integrated, using the concepts described herein.

12 Conclusions

In conclusion, the principles of IDA may be summarised as follows:

- The aim and boundaries of the system to be analysed, and of the analysis itself, must be clearly defined.

- The analysis must be founded on a complete understanding of the functions of the system and its components.

- At least a part of the analysis must cover all elements of the system. Passive components, such as those which only provide protection or support, should not be neglected.

- There is usually no single method which will achieve a complete analysis, so it is necessary to put together a set of methods appropriate to each task.

- Analysis should include a study of the system functioning as designed, as well as analysis of the causes, modes and effects of failures.

- The completeness of the analysis will be improved by using a combination of inductive (bottom-up) and deductive (top-down) methods.

- The completeness of the analysis will be improved by using the psychological map provided by a set of frames.

- The completeness and consistency of the analysis will both be improved by exploiting the logical relationships between the chosen analytical methods.

- The integrity of a safety (or any other) prediction depends on the continuing stability of the system.

Those principles should be used to guide decisions on whether, how and where IDA is to be applied.

IDA provides an auditable and traceable analysis, allowing people to use their judgement in selecting a set of methods to suit a particular task. It aims to extract the maximum value from the chosen methods. Above all, it is an aid to thinking.

References

[Bateson 1987] Bateson G: Steps to an Ecology of Mind. Jason Aaronson Inc. 1987. (Part III Chapter 2.)

[CFGM 1995] Chrysler Corporation, Ford Motor Company, General Motors Corporation: Potential Failure Mode and Effects Analysis (FMEA) Reference Manual. 2nd Edition 1995.

[Crawford 1999, 2000] Crawford J P K: What's Wrong with the Numbers? A Questioning Look at Probabilistic Risk Assessment. Parts 1 & 2. Safety Systems, September 1999 and January 2000.

[Crawford 2001] Crawford J P K: Some Ways of Improving Our Methods of Qualitative Safety Analysis and Why We Need Them, in Redmill F and Anderson T: (ed.) Aspects of Safety Management. Springer-Verlag, London, 2001.

[Cutler 1997] Cutler A N: Deming's Vision Applied to Probabilistic Risk Analysis. Second Edinburgh Conference on Risk: Analysis and Assessment. September 1997.

[Cutler 1999] Cutler A N: Normal Accidents: A Statistical Interpretation. Aegis Engineering Systems, Manchester, 1999.

[Deming 1993] Deming W E: The New Economics for Industry, Government, Education. Massachusetts Institute of Technology, 1993.

[Keynes 1921] Keynes J M: A Treatise on Probability. Macmillan, London, 1921. (Chapter XXXII)

[Kletz 1992] Kletz T: HAZOP and HAZAN - Identifying and Assessing Process Industry Hazards. Institution of Chemical Engineers, Rugby, 3rd edition 1992.

[Marshall 1999] Marshall S A: Personal communication to the author. 1999.

[Marshall 2000] Marshall S A: Presentation to the NDIA Fuze & Munitions Technology Symposium. Pleasanton, California, 10 April 2000.

[MOD 1996/1] Ordnance Board: Pillar Proceeding P120(1) Integrated Design Analysis for Fuzing System Safety. Ministry of Defence UK, 26 March 1996.

[MOD 1996/2] Defence Standard 00-56: Safety Management Requirements for Defence Systems, Part 1, Issue 2. Ministry of Defence UK, 13 December 1996.

[MOD 2000] Defence Standard 00-58: HAZOP Studies on Systems Containing Programmable Electronics, Parts 1 & 2, Issue 2. Ministry of Defence UK, 19 May 2000.

[Palmer 2001] Palmer S R: Personal communication to the author, 22 February 2001.

[Shewhart 1931] Shewhart W A: Economic Control of Quality of Manufactured Product. Van Nostrand, New York, 1931.

[Wheeler 1993] Wheeler D J: Understanding Variation - the Key to Managing Chaos. SPC Press, Knoxville, Tennessee, 1993.

Airport Risk Assessment: Examples, Models and Mitigations

John Spriggs

Safety Assurance Consultant, Roke Manor Research Limited,
Romsey, SO51 0ZN U.K.

Abstract

This paper addresses some of the hazards arising from an airport runway and its associated operations. It describes an approach to risk assessment in this context; some of the tools that may be applied; and some mitigations, both established and innovative.

Introduction

Some industries are only now making forays into the use of mission critical electronic systems. They have the luxury of starting with a clean sheet of paper. A lot of the tutorial material available on critical systems development seems to assume this clean start. In contrast, there are industries in which such systems have been in use so long that they have had to undergo radical replacement programmes as equipment wears out, or spare parts become unavailable. They experience a whole new set of problems involving the transition from the old system to the new, especially where continuity of service is paramount. There may also be hidden health hazards as materials now known to be hazardous may be present in the construction of the system to be decommissioned.

There is a third group – systems that have evolved steadily by the ad hoc introduction of electronic subsystems, for example, as demands have changed. Such systems are typically the subject of a plethora of prescriptive regulations, each originally produced in reaction to a newly identified hazard, or a particular accident.

One example of an evolving system is an airport. Some airports started as a shed on a flat field owned by an aviation pioneer or enthusiast, more were released by the military, others were originally associated with aircraft factories. Of course there are some airports that have relatively recently been built from scratch with the purpose of being an airport, like Denver International and Hong Kong International, but I believe that these are in the minority. Note, also, that both my examples were built as replacements for other airports that could not expand further, so they could still be regarded as evolutionary steps.

An airport is a multifunction distributed system that is part of a much larger system. You can think of it as being at the centre of a dynamic network made up of all the sources of cargo, passengers and the other people who travel to and from the airport; visitors, cleaners, et alia. But that is just the ground system; a large number of these networks are interconnected to form a huge communications network; the nodes are the airports with their hinterlands, and the dialogues are made up of aircraft.

This paper looks, as an example, at an extremely small part of that network, picks out a few hazards, introduces novel modelling tools that have been used in their analysis, and discusses some mitigating strategies. But first, here is a brief digression to revise the topic of risk assessment.

Risk Assessment

When the English poet William Cowper wrote "...*but to fly is safe*" he was not making the sort of unsupported assertion that engineers and service providers must avoid. Indeed, he was not even referring to aeroplanes; this was the eighteenth century after all [Cowper 1785]. A fuller quotation is:

> *"To combat may be glorious, and success*
> *Perhaps may crown us; but to fly is safe."*

Or, to put it the other way around, you could try to reduce risks in an activity by totally avoiding specified hazards, but in doing so you may miss out on greater benefits.

Like all modes of travel, flying is hazardous, but by identifying the inherent hazards and assessing the associate risks, we can put mitigating features in place so that the benefits can be achieved whilst the risks are reduced. A formal risk assessment carried out by the service providers, with support from equipment suppliers, and accepted by a Regulatory Authority, provides sufficient assurance for the general public to use the services without having to worry about doing their own risk assessments of the transport infrastructure before deciding how to travel.

A risk is "the combination of the probability, or frequency, of occurrence of a defined hazard and the magnitude of the consequences of the occurrence". This definition comes from the United Kingdom's Regulator of Air Traffic Services, the Safety Regulation Group of the Civil Aviation Authority, who defines a hazard as "a physical situation, often following from some initiating event, which can lead to an accident" [CAA 1998]. A risk is thus an attribute of a hazard.

Personally, I interpret the "combination" of parameters in this definition of risk to mean the formation of a two dimensional quantity. So, if the risk is to be reduced, it can either be done in the severity axis, or in the likelihood axis, or both.

To effect a decrease on both axes may be considered the best approach to risk reduction; but it is not always possible. For natural hazards such as an earthquake, typically we cannot do anything to reduce the likelihood, but there is much that can be done to reduce the consequences. For example, special building regulations can be put in place and the general public can be encouraged to keep "earthquake kits" to hand that contain emergency food, water, first aid materials clothing, etc. Alternatively, there is little that can be done to reduce the consequences of a mid-air collision of two aircraft, but there is much that can be done to reduce the chances of it happening. For example, the Air Traffic Control system is in place to monitor and maintain both vertical and horizontal separation; also, many aircraft have radar based collision avoidance systems installed.

The goals of risk assessment are:

- To derive the values of likelihood and severity of consequence for each hazard. These will, in general, not be precise values, but rather an informed judgement as to "order of magnitude".

- To use that information as a means of prioritising actions, i.e. which hazard requires the most work and so should be tackled first?

- To specify mitigating features as appropriate to each hazard; and

- To predict the effectiveness of those features in reducing the risk.

The last two points are usually extended to the specification of, and selection from, a number of mitigating strategies, possibly as part of a wider cost benefit analysis.

To stand any chance of achieving these goals we first need a list of hazards; a necessary precursor is thus hazard identification. When building a large system from a number of smaller ones we find that many of the hazards arise from the intra-system interfaces. When performing a risk assessment, then, we can start off by identifying those interfaces and the hazards arising from them. Where a system is made up of subsystems from different suppliers their "domains of influence" also need to be considered. I can identify and mitigate hazards in my domain; but I can only identify them in yours. [RIN 1998] The overall system owner needs to be able to coordinate and disseminate hazard identification information.

An airport has a lot of interfaces with the outside world, air traffic control has radio and telephones; there are navigational aids that communicate with aircraft, such as the distance measuring beacons and instrument landing systems; there are road links; there may be rail links; etc. We will consider an airside interface, the runway. It is "A defined rectangular area on a land aerodrome prepared for the landing and take-off of aircraft" [ICAO 1995]. It is the interface between the air navigation system and the ground handling area.

Some Collision Hazards and Their Mitigation

The runway is an interface; its air connections are in what is referred to as the Terminal Area of the airspace; they comprise the so-called STARs and SIDs, i.e. the Standard Arrival Routes and Standard Instrument Departures. The ground connectors are the taxiways, which have defined holding points where outgoing aircraft wait until they can use the runway.

The taxiways are themselves a mitigating feature. When airfields were literally fields it was normal practice to taxi directly from the parking position to the start point for the take-off run and, as there were no fixed routes or traffic rules, collisions occurred. The first such collision that resulted in a court case in the United Kingdom was reported in the Daily Mail newspaper of 5th October 1933. The defendant was Pauline Gower, who was later a director of the British Overseas Airways Corporation. The report of the case, quoted in the biography written by her son, shows that visibility from the cockpit of a tail-dragger aircraft was also a contributory factor. [Fahie 1995]

> *"... she ascertained that her passengers were alright, and started the engine. As she had just previously seen Mr Cubitt's machine on her right, she did not expect to have to look again. To look ahead she would have had to stand up."*

Of course, in the last seventy years or so of paved runway operations, many hazards have been identified and mitigations developed. In the main, these are embedded in prescriptive specifications, although recently Regulators have been moving over to a more objective setting rôle. [SSS 2001] This is intended, inter alia, to avoid the problem of the suppliers thinking that they have done enough when they have "ticked all the boxes". Some of the requirements were set many years ago. Things will have changed, there may be new hazards to address; some may have even gone away.

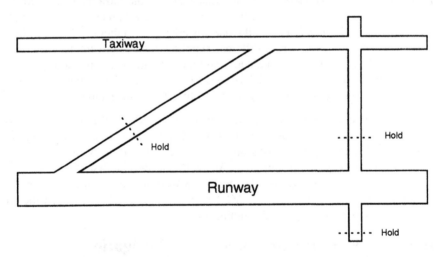

Figure 1 ~ Part of a Runway and Taxiway System Showing Holding Points

Having introduced taxiways, the regulations also needed to define holding points, also as a mitigating feature against collisions. Their purpose is to protect the arriving aircraft from those on the ground; it is clearly undesirable to have an aircraft, or other vehicle, crossing a runway whilst another is on final approach.

The specifications on where to put holding points are based on consideration not only of relative aircraft sizes, but also parameters such as the susceptibility to interference of the electronic landing aids.

There seem to be some ambiguities between the various requirements, for example those pertaining to aircraft approaches using the Instrument Landing System define notional surfaces above which obstacles, such as new buildings, cannot extend. However, these surfaces do not apply to mobile obstacles to the same degree; aircraft can wait at a point where a building of equivalent size would not be permitted. [Gleave 2000] Our risk assessment must consider not only the risk to landing aircraft due to those on the ground, but also the relationship in the opposite direction. What is the risk to an aircraft at a holding point due to landing aircraft?

One hazard is the impact of debris due to a landing accident. There is a model available in the public domain for deriving the probabilities of impact in such a situation. It is an empirical model developed by National Air Traffic Services Limited from publicly available accident data. It is documented in a report [NATS 1996] that was publicised in support of a 1997 Consultation Document of the UK Government's Department of the Environment, Transport and the Regions concerning Public Safety Zones around airports.

The model provides equations with which to determine either the impact or wreckage location of an aircraft following an accident. Several equations are described which cater for all permutations of:

- Aircraft operation (approach or departure);
- Crash from flight, or runway run off, and,
- Crash location (before or after the prepared runway surface).

These equations form a set of probability distribution functions of a crash occurring per unit area.

The probability of an aircraft being struck by debris due to a landing accident has four components, which are listed below. The product of these components gives the desired result.

p(Accident)	This figure represents the proportion of aircraft within the airport vicinity that have an incident that causes it to crash, or run-off the runway.
p(Accident Type)	This figure represents the type of accident the aircraft will have (i.e. crash short of runway on approach, or run off the side of the runway).
p(Risk Area)	This figure represents the region that will be affected by the resulting crash or run-off.
p(Aircraft)	This figure represents the probability of an aircraft being in the Risk Area when the accident occurs.

These four probabilities are derived from several sources. The Accident Probability, p(Accident), is set by the assessor based on data provided by the airport. A pessimistic example for this figure is taken to be one accident per million operations, i.e. p(Accident) = 1×10^{-6}. The worldwide average is quoted as 1.3 million departures per accident (of which about seventy percent are controlled flight into terrain), and in Europe, the rate is better than this. Figures from the Flight Safety Foundation [FSF 1999] give the fatal-accident rate involving Western built large jets in the full-member states of the European Joint Aviation Authorities (JAA) as over six million departures per fatal accident. We are interested in the non-fatal accidents too; the total European figure is of the order of two million departures per accident.

The Aircraft Probability, p(Aircraft), is also set by the assessor. It is dependent upon the airport environment, its operations and procedures. This parameter is the probability that an aircraft will be within the Risk Area at the time of an accident, and is comprised of a number of factors. It takes into account the time a single aircraft will spend in the Risk Area, the number of aircraft moving through the Risk Area and a Proportion of Risk Area factor.

The factor for Proportion of Risk Area is last is an artefact of the model, which requires a rectangular Risk Area to be specified. In practice the actual area of interest may not be rectangular, for example, consider a stretch of taxiway with a corner in it as shown in Figure 2. If the taxiway area of interest were between points 'A' and 'B', the Risk Area would have to be specified as shown. However, an aircraft cannot occupy a large section of the Risk Area (in normal circumstances), so Proportion of Risk Area provides a means to adjust the p(Aircraft) value to more accurately reflect the probability of an aircraft in the Risk Area.

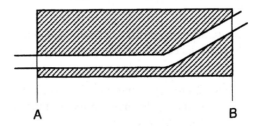

Figure 2: Proportion of Risk Area Example

Accident Type Probability and Risk Area Probability are independent of the airport under analysis. Both of these figures are taken from the crash location model equations. The Accident Type probabilities are fixed, set by the model components used. There are seven equations in the model that provide probability distributions for wreckage location following an accident in different scenarios. These equations are scaled in terms of probability per unit area. The Risk Area probability is determined by integrating the appropriate equations over the desired Risk Area limits.

We do not actually use the model to provide absolute numbers; rather it is used for comparisons of options. For example, we can ask questions like: "what if this holding point is moved back fifty metres, how does it affect the probability?" The model has been used "manually" calculating particular cases; but this gets very tedious and, hence, error prone if many options are tried. It was therefore also implemented in spreadsheet format, reducing the tedium to a data entry task. The spreadsheet functions were validated by comparison with the hand calculations.

More recently it has been implemented in a more flexible program using a windows based user interface. This also produces graphical output of results that can be overlaid on a model of the airport geometry that has been previously input via the keyboard, from CAD plans, or even as aerial photographs. The graphical output helps in the presentation of the options to non-technical stakeholders, such as the airport's senior management, so that they can make their decisions based on a visualisation of the scenario rather than "meaningless" raw numbers.

Care must be taken in the use of this tool; it will be necessary to justify the application of the underlying model to the particular circumstances under consideration. For example, the statistics upon which the model is based were collected from civil airliner accidents; they are unlikely to be applicable to the conceptually similar case of a military airfield operating fast jets that may take off and land in formation.

The tool may, of course, be applied to both existing and proposed airports. For example, an analysis may be performed considering areas of an existing airport based on actual usage data. Then, to support a safety case for airport expansion, the "differential risk" due to the projected traffic increase can be assessed. Similarly, new airport proposals may be assessed by performing "what if" analyses of candidate locations for taxiways, holding points and other assets, such as new buildings or navigational aids.

A "Credible Corruption" Hazard Mitigated by Stealth

Physical proximity to runways, etc., is not the only consideration when placing new buildings or navigational aids. A building acts as a reflector of electromagnetic energy, and so may cause multipath effects that affect navigational aids. An example hazard is "Multipath effects cause credible corruption to Instrument Landing System signals", i.e. aircraft performing automatic landings are being given inaccurate position data. This hazard can be a contributing factor in the landing accidents addressed in the last section.

In this case we can use a computer tool to model the postulated effects and then, iteratively, to design and evaluate any mitigating features that are needed. The bases of this model are the principles of physical optics and Maxwell's set of four equations relating electrical and magnetic effects. [Maxwell 1873] They are differential vector equations, which can be difficult to solve in non-trivial configurations. The tool that we have developed was originally designed to tackle the complexities of calculating the radar cross-section and coupling between closely positioned antennas at microwave frequencies. In this case, reactive and near field effects, combined with the presence of dielectric materials and complex geometries, prohibit the use of all but the most rigorous solution techniques.

Our tool is designed to be used as an adjunct to a popular Computer Aided Design system, and may also be used with others via graphics exchange formats. It provides the required rigorous solution by employing the finite-difference time-domain approach to solving electromagnetic problems, and as such is a general-purpose electromagnetic solver. It may thus also be used in electromagnetic radiation hazard calculations. Unlike conventional solvers that are restricted to problems described on cubic meshes, this tool can use rectangular and slightly non-orthogonal meshes to solve for complex geometries. It has been applied to the multipath problem on airports. For example, buildings at a major United Kingdom airport have been modelled; the predictions from the model were confirmed by measurements, and used as the basis of specification of features for multipath reduction.

As before, this tool may be applied to both existing and proposed airports. Existing airports may wish to explore the effects of a new building on the navigational aids; the tool can be used in trade-off studies to compare between position and/or configuration options. Alternatively, the airport operator may wish to install some new system, for example multi-static radar for monitoring the movements of vehicles on the airport. Multipath effects can cause multiple images of the same target; the tool can be used to design building features to minimise reflections, or to direct them away from the receivers. In effect, the tool is being used in the design of "stealthy" buildings.

Another Class of Runway Hazard

One significant set of hazards is that of obstacles in the path of an aircraft. As already noted, there are many regulations restricting the height and position of installations on an airport. Planners of developments outside the airport boundary are not always so constrained, although many jurisdictions have legislation requiring conspicuity lights on tall structures. In the context of runway operations, hazardous obstacles tend to be much smaller; we have briefly touched on the hazard of runway incursion by vehicles being mitigated by provision of holding points. These are augmented in practice with ground movement control procedures and traffic lights. Procedures have to be in place to cover normal use and failure situations, e.g. traffic should go on a green signal, but it should stop on red or no light at all.

The obstacle does not have to be something as large as a vehicle to be of concern. Debris on the runway can cause accidents. "Debris", which is from an Old French word meaning to break, implies parts of something broken. A lot of what may be found on a runway is in this category, for example a castor broken from a catering trolley, a part fallen off an aircraft, a detached runway light fitting, or part of the pavement itself. Not all the extraneous objects that may be struck by an aircraft are of this nature, however. There are misplaced tools or equipment; objects that have been blown by weather or jet blast; and, of course, some "debris" has fur, or feathers.

Foreign object damage due to debris on the pavement can occur anywhere in the aircraft manoeuvring areas of an airport. Damage can be caused by jet blast from one aircraft blowing debris onto another, or onto people, or onto logistics and infrastructure items. Damage can also occur directly to an aircraft striking debris on the pavement. A major event in which an aircraft sustains significant damage can cause delay to, or cancellation of, its flight, with the consequent knock-on effects of re-scheduling, increased workloads for airport and airline personnel, and loss of customer satisfaction. The major cost items will be repair and third-party liability claims. If debris is ingested by a jet engine, repair costs can exceed one million euro; the airline or its insurer will be looking to recover these costs.

On the active runway, the consequence of a foreign object damage event can be more severe than elsewhere on the airport due to the higher kinetic energies involved; it can even lead to loss of the aircraft, and consequent loss of life. Take off accidents are likely to be particularly severe due to the large fuel load. Such an event on the airport is likely to result in closure for a significant period, causing knock-on effects throughout the Air Navigation System of the region, which may increase the likelihood of further accidents elsewhere.

Debris on the runway is not a hazard in the formal sense; it is too loosely defined. There is a whole range of different accidents that may result from the same piece of debris. A stone on the runway may not impact an aircraft at all, or it may dent the skin, or it may get into an engine, and so on. This range of outcomes has an associated range of likelihoods. There is in fact a range of hazards. The hazards should be expressed in terms like "debris on the runway causes loss of aircraft". The severity is defined; the likelihood is a combination of sequential factors including the likelihood of actually striking an object, the likelihood of it causing damage of this severity, etc.

But what are the likelihoods; which are the highest risks? Unfortunately, there does not seem to be a common standard of data recording and sharing on what has been found where on airports, and what incidents, if any, resulted. Many operators have their own systems, but there are no statistics available combining these data.

The likelihood of striking an object and causing damage depends on, inter alia, size and position. For example, it could be expected that landing on an object would cause more damage than running into it during the deceleration phase. This suggests that debris removal operations should concentrate on the area upon which most landings occur. It could also be argued that it is here that debris fallen from aircraft is most likely to be found.

For take-offs, greatest damage would be expected at the point of greatest speed, i.e. around the point of lifting off, but this is a less well-defined area, depending upon aircraft type, loading and the weather. We could expect that the severity of the consequences of striking some debris to be related to the size of the item; a large stone can cause more damage than a small one. We can also assume that shape has an effect, a sharp object can cause more damage to a tyre, say, than a flat one.

These arguments seem reasonable; a small stone causes less damage than a larger item such as a misplaced tool and, considering its size, it is less likely to be struck. But consider the example, from August 1970, of a DC-8 Freighter taking off from New York JFK. It climbed to about four hundred feet, rolled twenty degrees to the left, crashed and caught fire. The investigation report cited the probable cause as "Loss of pitch control caused by the entrapment of a pointed, asphalt-covered, object between the leading edge of the right elevator and the right horizontal spar web access door in the aft part of the stabilizer."

Just because an event is very unlikely, it does not mean that it will not happen.

Existing and Proposed Mitigations

The best defence to the debris hazards is 'good housekeeping'. At present, the risk of foreign object damage is mitigated by procedures intended to make sure that things are not dropped, augmented by observation and 'sweeping' activities. There will be a schedule of inspection activities to be carried out by the Operations Department. These inspections are not just to find debris; there are also requirements to check for damage of, or contaminants on, the pavement. Additional inspections will be carried out when prompted by air traffic control, who will be either passing on a pilot's observations of debris, or requesting an additional sweep when an aircraft has landed after declaring an emergency, for example.

However, the area where the risk is highest, the active runway of a busy airport, cannot be swept on a periodic and frequent basis without reducing the number of aircraft movements, i.e. adversely affecting airport capacity. It is not only capacity that can be affected; there was a recent safety incident reported in which the driver of an inspection vehicle had to take avoiding action when he saw an aircraft landing behind him.

We are investigating what is in effect an on-condition maintenance concept for runways. Under such a concept, periodic sweeping activities would be less frequent, but the runway would be continuously monitored by automatic sensors to warn of the appearance of significant pieces of debris. This places Integrity requirements upon the sensors; they must not generate many false alarms but, of course, they are also required to detect all significant items of debris. Continuity of Service is also an important parameter; the sensors are required to operate all the time that the airport is required to operate, e.g. at night and in severe weather conditions.

Even though the Integrity and Continuity of Service requirements have yet to be specified in detail, it is thought unlikely that a single sensor type can be found to fulfil them. This is because, apart from the need to operate continuously whilst minimising false alarms, it is a difficult sensing task. "Debris" is not a neatly defined set of objects to sense. Different sensors excel in different conditions; their outputs need to be combined to give adequate performance over all expected conditions and types of object.

One approach is to use one sensor type for basic surveillance and, when it raises an alarm, task a different type of sensor to check the findings of the first. For example a scanning millimetre-wave radar sensor could, on detecting a small object on the runway, direct an automated visual sensor to the place and, if that also detects debris, an alarm would be raised for the duty officer in Operations to view the images to decide what, if anything, needs to be done. Note that three sensor systems are actually involved here, the duty officer being the third.

An alternative is to combine the outputs of a number of diverse sensor types to form the quantity to be checked for alarm conditions. The combination process not only needs to take into account the sensor declarations, but also any associated position information. For example, if all the sensors are giving out low confidence indications of debris, but agree on the location, there is likely to be something there, albeit small. If one sensor gives a strong response at a location and the others report nothing, it may be something benign to which the first sensor is particularly sensitive, like a quarter wavelength sized piece of metalised foil. If a large number of detection events are made all over the runway, it is likely to be a weather phenomenon, like a hailstorm.

To take best advantage of the position information it is necessary to ascertain for each point in the area of interest, and for each sensor, the probabilities of false positives or false negatives and, where a sensor can correctly identify that there is a piece of debris, the distribution of location accuracy. Some sensors may be very good at detecting the existence of debris, but poor at specifying the exact location, others may be the other way around. Note that the distributions may be functions of position, e.g. many sensors will be more accurate at close range.

Each sensor can now be used to produce a map of the whole area of interest with the probability of the presence of debris at every point. All these maps may be combined into one map by using one of a number of established techniques, including Bayesian Filtering; Neural Networks; or Fuzzy Logic based Expert Systems. Unfortunately all of these techniques have been over-hyped at one time or another, none is the universal panacea that some people perceive it to be.

However, theses fusion techniques do work, given appropriate initial training data, and good quality sensor data. What would be produced is a single map with an indication for the entire coverage area of the presence or absence of debris. These map data would then be compared with pre-defined thresholds to raise an alarm. The thresholds could be defined to be different in different areas so as to increase sensitivity in areas of greatest threat. They could also be adjusted for different ambient conditions, visibility, for example.

A fixed sensor installation has been implied, but this need not be the case; the sensors do not have to be mounted on posts or embedded in the runway. They can be mounted on an inspection vehicle to increase the effectiveness of its operation. Indeed, this may prove the most cost-effective solution for some airports.

Conclusions

This paper has considered risk assessment applied to airport runways. Only a small proportion of the many types of hazard that can be encountered there have been considered. Some novel computer-based tools have been described that support the assessment by helping in the specification and evaluation of mitigating features.

In the introductory section it was noted that an airport is an evolving system, with new subsystems being introduced as necessary to address new problems. The newest candidate for such a subsystem was discussed. Debris on the runway is not in fact a new problem, but it has only recently become a practical proposition to deploy an automated detection system.

It has been suggested that a future solution to the debris problem could be to automate the inspection and collection process by using robots to identify and pick up hazardous debris on the runway. Now, that would be an interesting safety problem! In such a new application domain, we could have the luxury of starting with a clean sheet of paper after all.

Acknowledgments

The debris detection work presented in this paper comprises aspects of a study carried out for EUROCONTROL, the European Organisation for the Safety of Air Navigation, under contract C/1.126/HQ/EC/01, together with work previously carried out by Roke Manor Research Limited.

Some of the techniques mentioned are the subject of patents held by Roke Manor Research Limited.

References

[CAA 1998]   Civil Aviation Authority: CAP 670: Air Traffic Services Safety Requirements, CAA London, 1998.

[Cowper 1785]   Cowper W: The Task, Book. Three, 'The Garden', 1785

[Fahie 1995]   Fahie M: A Harvest of Memories ~ The Life of Pauline Gower M.B.E., GMS Enterprises, 1995

[FSF 1999]   Flight Safety Foundation: Flight Safety Digest Special Report ~ Killers in Aviation, November 1998 → February 1999

[Gleave 2000]   Gleave D: Risk Management and Legal Responsibility for Instrument Flight Procedure Designers, presented at International Procedures Design Congress, Versailles, France, October 25th 2000

[ICAO 1995]   International Civil Aviation Organization: International Standards and Recommended Practices ~ Aerodromes; Annex 14 to the Convention on International Civil Aviation, Volume 1: Aerodrome Design and Operations, Second Edition. ICAO Montreal, 1995.

[Maxwell 1873]   Maxwell J C: Treatise on Electricity and Magnetism.

[NATS 1996]   Cowell P G, Gerrard R, and Paterson D S: A Crash Location Model for Use in the Vicinity of Airports, R&D Report 9705, National Air Traffic Services, 1996.

[RIN 1998]   Spriggs J, Benstead P: Risk Classifications Schemes for Satellite Navigation, in the Proceedings of NAV98, Royal Institute of Navigation, 1998

[SSS 2001]   Penny J, Eaton A, Bishop P and Bloomfield R: The Practicalities of Goal-Based Safety Regulation, in Redmill F and Anderson T: Aspects of Safety Management, Proceedings of the Ninth Safety-critical Systems Symposium, Springer-Verlag, 2001

Communicating Risk: Reconfiguring Expert-Lay Relations

Gabe Mythen
Manchester Business School

1. Introduction

In contemporary culture, risk has become a ubiquitous issue, casting its spectre over a wide range of practices and experiences. Despite such omnipresence, the meaning of risk is inherently uncertain and contestable. Since the Enlightenment period, prevalent social bodies have sought to accumulate information about the nature of risk. Without doubt, this process has facilitated heightened risk awareness within institutions and improved risk consciousness amongst individuals. In contemporary society, risk issues such as food safety, biotechnology and international terrorism are currently being debated by politicians, scientists, academics and the general public. Nonetheless, the growing public debate about risk and the advancement of scientific knowledge have not led to public perceptions of a safe and secure environment [Pidgeon 2000]. Somewhat paradoxically, as the 'answers' to risk dilemmas are uncovered, more complex questions are generated. Thus, it would appear that the Faustian bargain for knowledge about risk is an increase in uncertainty within everyday life. In this climate of widespread indeterminacy, the issue of how risks are communicated has become a focal concern. In Britain, academic, media and public interest in risk has been accentuated by a series of high-profile governmental communication failures.[1] At a structural level, acute deficiencies in information presentation and a lack of attention to the hermeneutic process have highlighted the absence of coherent communications strategies within several risk regulating institutions.

In this paper, I wish to construct a critique of dominant institutional methods of risk communication and to explore the potentialities of a more holistic approach to risk communications. Rather than being perceived as an end in itself, I contend that risk communications should be viewed as an articulation point through which the definition, assessment and regulation of risk flows. In the first half of the paper, I trace the traditional divide between the public and experts, unpacking the evolution of two competing languages of risk. In fleshing out the hiatus between expert systems and lay individuals, I employ the Bovine Spongiform Encephalopathy (BSE) crisis as heuristic. Subsequently, I go on to assess the general social learning which can be drawn from this particular case. Finally, I revisit the expert-lay divide, arguing in favour of a reconfiguration of the risk communications process from one of risk education to one of risk negotiation. Here, I seek to illuminate the

---

[1] The Salmonella affair and the BSE crisis being the most salient examples.

fluidity of expert-lay relations and to offer some tentative suggestions for a remodelling of risk communications to promote dialogic debate between stakeholders.

2. The Communications Divide: Theorising Expert-Lay Difference

At a rudimentary level, risk communication can be seen as a defining moment in the social construction of risk. Various communications about risk will inform and shape individual and group perceptions, which, in turn, influence values and behaviour.[2] As far as the diverse stakeholders influenced and affected by risks are concerned, 'getting one's point across' is central to involvement in the definition and evaluation of risk.

But what exactly do we **mean** when we talk about risk communication? In its simplest form, risk communication can be understood as a language of probabilities which provides an estimation of the likelihood that a particular hazard will result in harm or damage. Stripped bare, risk communication is 'the label used to refer both to the content of any message concerning a hazard and the means of delivering that message' [Breakwell 2000: 110].

On the basis of this definition, classic studies in the field of risk research have demonstrated discrepancies between the information disseminated by technical experts and the interpretations of these messages by the general public [Covello 1983; Slovic 1987]. Thus, it has been widely acknowledged that public perceptions of risk do not always tally with the informed opinions of scientific, technical and political experts. Within risk communications literature, this acknowledgement has led to the 'scientific' or 'expert' perspective being contrasted with the 'lay' or 'public' perspective [Lupton 1999; Macintyre et al, 1998]. Traditionally, these two perspectives have been conceived as opposite poles of the compass. On the one hand, the public are accredited with forming 'subjective' interpretations of risk that are influenced by social networks, emotions and fear. On the other, experts are accredited with the 'objectivity' provided by scientific investigation and statistical principles. Whilst risk communicators have understandably favoured the scientific-objectivist approach [Renn 1998: 5; Beck 1992], the general public have been found to judge risks with reference to a markedly different set of criteria [Langford et al 1999: 33; Lupton 1999].

These contrasting styles of sense making have encouraged academic researchers to home in on lay misperceptions of risk. Hence, the general public have been found to amplify distant hazards which are unlikely to materialize, whilst overlooking more profane local dangers [Flynn & Slovic 1995; Kasperson 1992]. Similarly, the

---

[2] For example, in Britain, egg sales halved following Edwina Currie's maladroit comments about the risk of contracting salmonella. More recently, the BSE crisis has had a marked impact on beef consumption at home and abroad. See, Macintyre et al [1998: 228] and Smith [1997].

public have been charged with overestimating 'man-made' threats and underestimating hazards produced by nature [Bennett 1998: 6]. These - and other - seemingly inaccurate perceptions of risk have been explained with reference to either public ignorance or ineffective risk communications [Macintyre *et al* 1998: 230; Fisk 1999: 133].

Unfortunately, many studies which have sought to emphasize public misperceptions have slavishly focused on the dissemination of the message and unwittingly reproduced a one-way model of communications. What has not been at issue – at least until relatively recently - is the manner in which experts gather, interpret and represent information about risk. Traditionally, institutional methods of risk communication have tended to focus on purveying statistical estimations of harm parcelled up in scientific and technical language. Whilst in some ways this is understandable - without quantitative predictors risk cannot be adequately framed - such an approach has historically neglected the rationale underpinning public fears and overlooked the situated cultural context in which risk communications are 'made to mean'. Thus, risk communication has been treated as if it took place in an ideological vacuum. This unsophisticated approach has served to abstract individuals from the situated everyday contexts in which information about risk is routinely encountered.

Challenging the dominant trajectory of institutional communications, here I contend that differences in the content and interpretation of risk ought to lead us away from exercises which pinpoint 'aberrant' public perceptions and toward a closer examination of the broad range of influences on sense-making. Contra the public misperceptions approach, what is required is a more nuanced and inclusive understanding of the risk communications process. By factoring social and political aspects in to our definition, we can recast risk communications as a set of interactions through which society identifies, quantifies and regulates risk. If we employ this wider understanding, the process of risk communications invites in a spectrum of stakeholders [Powell & Leiss 1997: 33; Weyman & Kelly 1999: 24]. According to this wider definition, there is a need for risk communications to address both individual and collective understandings of risk **in addition to** scientific and statistical estimations of harm.

3. The Two Languages of Risk

In the restricted definition, the central objective of risk communication has been to increase individual and social safety by conveying information about hazards. Hence, the guiding aim of institutional risk communications has been to educate and persuade the public to make rational choices and decisions about risk. This objective has manifested itself in various attempts to use scientific information to raise public consciousness and to modify practices or behaviours associated with risk. The desire to translate scientific understandings into acceptable public language is indicative of a long-standing hierarchy in risk communication [Krimsky & Golding 1992]. At a macro level, the process of risk communications can be recast as a power game, where experts and major stakeholders such as government and industry have held the chips [Irwin 1995: 52-53]. Risk communication, under its traditional conception, meant power over others constituted by persuasive messages based on a pervasive worldview. Under the traditional model of risk communication the ultimate objective for experts was to persuade the public of the accuracy of objective knowledge and facts:

> The underlying assumption was that if the average person received enough information about a particular risk, he or she would respond logically, be persuaded, and act in accord with the scientific view. In this definition of risk communication, the message about risk travelled only one way – from the scientific and governmental communities to a largely uninformed public where it would be understood and accepted in a somewhat uniform fashion [Valenti & Wilkins 1995: 179]

This linear model of risk communications understands science as the arbiter of rationality and truth and supposes citizens to be ill informed. Under this model of risk communications, ideological and linguistic divides have inevitably emerged.

Technically speaking, in order to communicate, experts and lay actors need to refer to a set of signs and symbols. To actively engage through language participants must share the same signs and symbols. Regrettably, this basic rule of communication has often been overlooked by risk practitioners. What makes perfect sense in the language of scientific rationality may literally be non-sense to those using everyday social discourse [Beck 1995]. The rationality clash between expert and lay perspectives has driven risk communications literature to draw a theoretical wedge between two forms of discourse: that of the 'expert' which is grounded in scientific, specialized and statistical knowledge, and that of the 'public' which is grounded in social and intuitive knowledge. From this, it is supposed that there are two conflicting assessments of risk simultaneously in circulation. In this *oeuvre*, risk communication has famously been described as a dialogue between actors who speak different languages [Figure 1].

'Expert' Assessment of Risk	'Public' Assessment of Risk
Scientific	*Intuitive*
Probabilistic	*Yes/No*
Acceptable risk	*Safety*
Changing knowledge	*Is it, or isn't it?*
Comparative risk	*Discrete events*
Population averages	*Personal consequences*
A death is a death	*It matters how we die*

Figure 1: Characteristics of the two languages of risk communication [Powell & Leiss 1997: 27]

As Powell & Leiss infer, the language used by experts is often couched in quantitative or scientific terms and is founded on an approach to risk assessment that evaluates populations. In contrast, the public are principally concerned with effects on individuals. Thus, whilst risk regulators may aim to protect public health on the basis of numbers and averages, lay individuals will perceive risk in terms of personal injury [Fisk 1999: 134]. Both the public and experts are seen to be employing probability indicators. However, the reference point differs between the population as a whole and the individual within that population. Similarly, as Langford *et al* [1999: 34] note, scientific experts and government spokespeople have traditionally centred on an absence of negative results, whilst the public prefer to be reassured about positive safety. Whilst governmental assessments are tied to actuarial principles (no risk unless proven otherwise), the public generally favour the precautionary principle (possible risk unless proven otherwise). The apparent incommensurability of public and expert assessments suggests that - in their purest forms, at least - the two paradigms are incompatible:

> The criteria of proof for a scientist and a layman are different, particularly in a crisis. The scientist will stick to notions of scientific certitude since to depart from these would seem to negate his or her professional integrity: for the public, probability or even possibility is sufficient proof [Harris & O' Shaughnessy 1997: 35].

It is such fundamental inconsistencies of language and method which have worked to produce different values and attitudes toward risk.

4. Miscommunications: Learning from the BSE Crisis

Historically, the expert voice of science has been used to give authority to risk communications. Of course, without reference to the natural sciences, identifying and quantifying risk becomes an impossible task. On the basis of this alone, we should challenge dogmatic thinking which casts science and technology as tools of mass deception. On the contrary, there is nothing inherently 'bad' about science and/or technology. However, we must earnestly recognise that, in addition to alleviating the deleterious effects of health risks, science and technology are also manufacturers of risks.[3] As far as risk communications is concerned, it is paramount that science is presented in a precise and balanced manner. In certain instances, accurate evidence gathered by scientists and technical experts has been spun and re-packaged to construct misleading information about risk.

One of the most notorious examples of the misuse of science in risk communications is the Conservative government's handling of the BSE crisis in

---

[3] Nuclear, chemical and genetic technologies being obvious examples.

Britain.[4] In précis, despite being furnished with a range of information indicating a potential risk to public health, the government consistently refuted a link between BSE in cattle and a new variant of Creutzfeldt-Jakob Disease (nvCJD) in humans. Following the first wave of public concern about the possible transmission of BSE from cattle to humans in the late 1980s, the government - fearful of a loss of consumer confidence and a subsequent fall in export profits - recruited scientific experts to play down the possible connection between BSE in cows and nvCJD in humans. In an exceptionally distasteful incident, the Environment Minister John Gummer was filmed feeding beef-burgers to his four year-old daughter, Cordelia. This was, of course, a rather desperate attempt to allay rising fears about the risk of eating British beef and to inveigle the public into further consumption. As discussed by respondents in Reilly's study, the blanket denial of potential risks by senior politicians was particularly disingenuous [Reilly, 1998: 135]. Given the scientific information available to government ministers at the time, the conclusion that eating British beef presented no risk whatsoever to public health was either grossly incompetent or simply mendacious.

The aftermath of the BSE crisis has brought into acute focus the dangers of suppressing subaltern discourses and foreclosing vital scientific debate. In the early stages of the BSE crisis, the government failed to clarify the potential risk to public health and instead hid behind a screen of scientific rhetoric.[5] To be fair, it is quite possible that the government believed that the public would more willingly accept the neutral language of science, than the interested voice of government. Further, those involved in risk communications were not helped by the indeterminacy of scientific information about BSE. Nevertheless, when faced by a dearth of definitive facts, the government unwisely opted to couch the crisis in technical terms.

As the crisis played itself out, whilst the government continued to impart technical information, the issue was progressively framed 'through a series of impulsive, rhetorical and symbolic acts which gave the crisis meaning' [Harris and O'Shaughnessy, 1997: 30]. Throughout the BSE saga, the role of media symbols in influencing public understandings of risk was not properly appreciated by the British government. At the same time as the government were offering dry scientific refutations of a move across the species barrier, public opinions became influenced by a disturbing set of iconic images. Footage of stumbling cows were played and replayed on news bulletins and graphic images of burning cattle carcasses routinely appeared in newspapers. Of course, public perceptions of risk cannot simply be 'read-off' from media reports. Nonetheless, the symbolic representation of BSE does appear to have played an important role in shaping public attitudes [Adam, 2000].

---

[4] A detailed analysis of the BSE crisis is beyond the ambit of this paper. For solid reviews, see Ratzan [1998] or Hinchclffe [2000].

[5] For example, the potential causes of BSE were rarely explained in an accessible fashion. Later in the crisis, the terms 'BSE', 'mad cow disease' and 'nvCJD' were often conflated by politicians.

Although the government clearly failed to appreciate the power of the media in setting the public agenda, media outlets must themselves exercise due responsibility in representing risk. In the media scramble to name and blame, the right of the public to an informed and considered range of risk information must be upheld. Indeed, media presentation of an appropriate range of information is vital to everyday decision-making and risk avoidance strategies. In many respects, the BSE crisis serves as a prime example of the complex and highly charged terrain of risk communications in a multi-media age. But which specific lessons can be learnt from the BSE communications failure?

Firstly, the BSE imbroglio indicates that overemphasizing the technical/scientific dimensions of risk can serve to alienate the public and may foster distrust in expert institutions. Several studies have indicated that, during the crisis, the general public perceived prominent politicians to be over-confident in the ability of science to protect public health [Harris & O'Shaughnessy, 1997; Reilly, 1998]. At a deeper level, senior politicians were relying upon a dangerous cocktail of absolute trust in science and fingers-crossed fatalism. Post BSE, the evidential basis of the bifurcation between scientific experts and the public has at least been acknowledged in risk communication guidelines:

> Most of the time - though not invariably - scientists will accept
> the existence of a causal link only once there is good evidence
> for it. Until then, links are 'provisionally rejected'. The lay view
> is much more likely to entertain a link that seems intuitively
> plausible, and reject it - if at all - only if there is strong evidence
> against it [Bennett 1998: 14].

Identifying and understanding these underlying value conflicts may be an important step forward. However, actually integrating seemingly binary oppositions into a meaningful dialogue about risk presents a more complex challenge.

Secondly, wider acknowledgement that science and technology are not always able to provide precise estimations of harm is required within expert circles. In the aftermath of BSE, there is increasing recognition in the public sphere that science is a contested sphere and should not be expected to provide immutable truths. Scientific expertise is now seen as fallible and there is greater awareness that empirical findings can be ambiguous. In an area so fraught with uncertainties, the 'science of the matter' will remain difficult to configure and evaluate. Undoubtedly, the BSE crisis has acted as a watershed in terms of the relationship between experts and the public. Risk communicators need to be aware that public trust in institutional estimations of risk has declined markedly in recent years [ESRC Report 1999]. This said, the broader trend of questioning expert systems does not indicate that the public should or ought to dispense with science. On the contrary, imperceptible risks such as nvCJD demonstrate that society has become **more** rather than **less** dependent on science. Again this raises the stakes for risk assessors and places increased pressure on scientific agencies and regulatory bodies.

Thirdly, the BSE case illustrates that - however useful as a heuristic device – the bifurcation between 'two languages of risk' exaggerates difference and accentuates perceived divides between experts and the public. Whilst there is some justification for mapping a **general** process of polarisation, in its purest form, the lay-expert language divide reproduces unhelpful stereotypes. It should be noted that - throughout the duration of the BSE crisis - there was a fair bit of hybridity in evidence, both in terms of risk assessment and communication strategies. For example, cabinet ministers made frequent emotive appeals to nationalist sentiment, whilst members of the public took issue with the government over the science of the issue. When a Siamese cat in Bristol contracted BSE, many members of the public felt that this was evidence enough that the disease could cross the species barrier, despite the continuing denials of experts [Harris & O'Shaughnessy 1997: 33]. In a similar vein, recent research indicates that the public considered feeding herbivores on meat derivatives a scientifically unsound activity [Cragg Ross Dawson Report 2000]. In the shadow of BSE, experts must accept that sometimes 'the public is right' in its estimations of harm [Fisk 1999: 133].

With these examples in mind, we must be wary of perceiving risk communication as a process of translating danger from one language into another. In so doing, we reduce risk communication to risk education, misrepresent 'expert' scientific knowledge as objective and render the public unable to cope with expert assessments. Instead of assuming that scientific discourse has validity whilst public knowledge is bereft of worth, we need to move towards a model of risk communications in which the shortcomings and the utility of both forms of knowledge are properly appreciated. Further, we must recognise that conflicts take place **within** as well as between expert and lay groups [Wynne 1996]. In the BSE case, experts representing science, government and industry differed in their opinions as to the probability of contracting nvCJD from British Beef [Hinchcliffe 2000: 144]. In attempting to draw out the generalities present in expert and public perceptions of risk it is prudent to be sensitive to the diversity of opinion inside expert and lay groups [Caplan 2000]. Whilst it is widely accepted that the aspirations of the public cannot be adequately understood as an amorphous mass, recognition of conflict within expert circles is not always as forthcoming.

Fourthly, as the Phillips report[6] indicates, risk practitioners need to radically rethink strategies for communicating uncertainty. The human and social cost of communicating 'certainty' in a chronically uncertain situation has been high. Had the government not presented scientifically uncertain information as certain, the public and the families affected by nvCJD might have been more forgiving. The ability of the public to handle complex and uncertain risk information was poorly acknowledged by the Conservative government in the BSE crisis [Wynne 1996a].

---

[6] The full BSE inquiry report is available at http://62.189.42.105/report/toc.htm.

In sharp contrast, empirical studies have upheld the capacity of citizens to handle uncertain and diverse information about risks [CESC Report 1997; ESRC Report 1999; Macintyre *et al* 1998]. With direct reference to information about BSE, Reilly [1998: 135] posits that 'respondents...were quite aware and appreciated that decisions they made were rarely based on absolute certainties, but rather a balance of probabilities. Clearly, some members of the public are better informed about scientific processes than has traditionally been assumed. In future, greater intelligence should be credited to the public and - if present - expert acknowledgement of uncertainty should be forthcoming. Politicians, civil servants and scientists must demonstrate more confidence in the maturity of modern citizens to share the burden of uncertainty.

Despite nascent evidence of lay skills in dealing with complex and uncertain information about risk there has been patchy acknowledgement of public sophistication in policy circles [ESRC Report 1999: 6]. It needs to be recognised that risk misperceptions apply to experts, journalists and policy makers as well as public stakeholders [Bennett 1999: 4]. Complete objectivity is an impossible goal for risk assessors as well as the public. Scientific experts do not inhabit a vacuum and they too will be affected by emotional and psychological influences. As Flynn and Slovic note [1995: 334], the various stages of the risk assessment procedure cannot be immunised against subjectivity. Scientists, technical experts and risk assessors do not inhabit a value-free space and will be influenced by wider social and cultural factors in their assumptions and duties. Conversely, whilst the public have been found to be more inclusive of social and cultural considerations, lay understandings of risk are not exclusively governed by habit, experience or subjective thought. All of this implies that we need to appreciate that expert and lay categories are contingent and open to fluctuation [Crook 1999]. Outside of their employment, 'experts' are at once 'laymen', 'consumers' and the 'public'. Consequently, the distinction between 'the public' and 'the experts' is not as clear-cut as has traditionally been assumed [Tennant 1997: 149].

So, have risk communication practices improved in the light of the BSE imbroglio? Well, yes and no. On the positive side, risk communicators are beginning to realise that 'absence of evidence is not the same thing as evidence of absence' [ESRC Report 1999: 7]. In addition, there is a growing awareness that scientific judgements on risks and uncertainties are inevitably influenced by subjective assumptions. Promisingly, recognition that framing assumptions will affect the structure and results of risk assessments is increasingly evident in institutional policy documents [Coles 1999: 196]. Concomitantly, evidence suggests that a more reflective and socially aware understanding of risk is being developed within expert institutions.[7] Certainly, there is now general acceptance of the multi-dimensionality of risk (Jungermann 1997: 141). As the echoes of BSE reverberate

---

[7]For example, the Food Standards Agency's *Policy on Openness* states that the agency will act on the precautionary principle in matters of food health and safety. For further details, visit www.foodstandardsagency.gov.uk.

around society, state institutions have acknowledged the need to be sensitive to public opinion. Most practitioners are now agreed that greater attention must be directed to the interests and aspirations of diverse stakeholder groups [Dean 1999: 144; Handmer 1995: 91]. Whether these developments are indicative of a delicately fashioned veneer or a franker commitment to openness and accountability remains to be seen.

Those involved in public safety are now aware of the different windows through which individuals interpret risk and attempts have been made to actively involve the public in the risk communications process.[8] At a wider level, a range of public involvement vehicles such as citizens' juries and deliberative opinion polls are being advocated as potential mechanisms for facilitating dialogical risk communication [Pidgeon 2000]. However, in spite of this apparent shift in the trajectory of risk communications, our ingrained views of what counts as 'expert knowledge' seem harder to shift. Traditional conceptions of risk communication as a means by which experts educate an ignorant public remain hard to dislodge, even within the framework of apparently progressive forums of communication, such as consensus conferencing [Purdue 1995]. Innovative attempts to build bridges between stakeholders must be commended, but we must ensure that public participation is not simply 'tagged-on' after technical and scientific experts have already set the agenda [Irwin 1995: 79].

5. The Multi-dimensional Future of Risk Communications?

Fischhoff has famously traced the evolutionary process of risk communication, from simply 'getting the numbers right' to involving diverse stakeholder groups in the process [Figure 2]. In theoretical terms, the maturation of risk communication can be conceived as a subterranean shift in the conditions in which the various stakeholders encounter each other during risk dialogue. At a practical level, recent risk communications episodes indicate that there is much scope for evolution.

In this section, I wish to argue in favour of the dynamic development of the risk communication, from a process of 'risk education' to one of 'risk negotiation'. To this end, I explore the implications of possible transformations in risk communications for the relationships between various stakeholders involved in the process. Having established that expert-lay categories are ideologically restrictive, I wish to conclude by arguing for the reconfiguration of the risk communications process as a means of bringing in a diverse range of stakeholders and improving the quality of risk dialogue.

---

[8]Companies such as Unilever and Nirex have championed public involvement and developed avenues for stakeholder dialogue about risk.

> **The Evolution of Risk Communication**
>
> - All we have to do is get the numbers right.
> - All we have to do is tell them the numbers.
> - All we have to do is explain what we mean by the numbers.
> - All we have to do is show them that they've accepted similar risks in the past.
> - All we have to do is show them that it's a good deal for them.
> - All we have to do is treat them nice.
> - All we have to do is make them partners.
> - All of the above.

Figure 2:The various prescriptions for communicating risks to the public [Fischhoff 1995: 138]

Historically, the conception of risk communication has evolved 'from an emphasis on public misperceptions of risk, which tended to treat all deviations from expert estimates as products of ignorance or stupidity, via empirical investigation of what actually does cause concern and why, to approaches which promote risk communication as a two-way process in which both 'expert' and 'lay' perspectives should inform each other' [Bennett 1998: 3]. Taking this argument a step further, the current challenge is to move towards a multi-dimensional notion of risk communications by developing new conduits through which diverse stakeholders can input into the process. This potential for advance can be perceived as part of a broader shift in the wider context in which various stakeholders encounter each other, from a relationship based on paternalism to one of partnership. The development of a holistic approach in risk communications may ultimately serve to facilitate understanding of the common ground between stakeholders implicated in the production, definition and regulation of risk.

As Fischhoff's evolutionary plot infers, the risk communications process is inherently political. Risk communication is - and will remain - an articulation point for competing and conflicting values and assumptions. Thus, effective risk communications must take due account of the motivations and values which frame risk understandings, rather than simply striving to 'get the numbers right'. The sense which is made out of risk communications will be mediated by social and cultural identities of class, gender, age, ethnicity and political affiliation. This indicates that risk communication should not be seen as a panacea:

> One cannot expect to quiet raging controversy with a few hastily prepared messages. One cannot assume that expensively produced communications will work without technically competent evaluations. Those who ignore these issues may be the problem as much as the risk is [Fischhoff 1995: 144].

Further, the meaning of risk in society can be expected to emerge and evolve in the course of social and political debates. We should therefore understand risk dialogue as the negotiation of difference and an attempt to reach agreement out of a plurality of competing perspectives. Rather than denying or attempting to circumnavigate risk conflicts, we should instead accept that different opinions will emerge from **different**, not unfounded or illogical standpoints. Growing the conditions in which relations of mutual respect and trust may be fostered in risk communication means accepting that competing understandings of risk are expectable and healthy.

This is not to naively assume that risk assessment procedures can universally welcome-in stakeholder participation. Of course, this would be technically impractical and may lead to inertia at the level of regulatory decision-making. However, we do need to think creatively about how stakeholder groups can be represented at the various nodes of the risk communications circuit. In particular, the formative stages of risk definition should be seen as a site which is ripe for reconfiguration. Previous risk communications failures have been exacerbated by a refusal to engage in early consultations with the public and a reluctance to involve diverse stakeholder groups into the decision-making process.

At the risk of repetition, requesting the integration of different value-sets into the process should not be misconstrued as a demand for the removal of scientific rationality from the risk communications equation. On the contrary, accurate science remains absolutely crucial in determining risks to individuals and societies and quantifying these risks in communicative terms. However, science should be treated as organic and contingent and must not be seen as the **only thing** to be attended to in risk communications. Contra the established tradition, it is vital that the transmission of technical information by experts is not seen as a terminus:

> Communicating risks is not just a matter of ensuring that one's messages are delivered and listened to. It is also very much a process of empowering individuals and communities to appreciate different types of viewpoints about risks (including technical risk assessments), scrutinize beliefs and perceptions about risks, and sharpen the skills necessary to make balanced judgements on risks which consequently impact on individuals or communities [Scherer 1991: 114].

Taking public concerns seriously means, in the first instance, listening to the expression of those concerns with an open mind. It is only after acknowledging public concerns that we can start to talk to each other, rather than past each other in debates about risk [Frewer *et al* 1999: 21]. Although recent risk communication guidelines support the ideal of risk communication as a democratic process of negotiation [ILGRA Report 1998; FAO/WHO Report 1998], we should not be naïve. In order to actualise this understanding of risk communication there are many obstacles to surmount. Risk communication remains a contested terrain and we should recognise that certain stakeholders may 'play dirty' [Leiss 1995: 687]. Inevitably, risk communications will remain pregnant with strategies of blame, obfuscation and self-interest. Nevertheless, such proclivities are more malleable and surmountable if they are situated within a framework which is open and fair. The goal of contemporary risk communication should be the facilitation of public dialogic arenas where power and strategy are discouraged and fairness and deliberation are encouraged [Wales & Mythen 2002]. Even supposing that risk communications can be reconceived as a mode of democratic dialogue, for some, the objective of this dialogue may still be coercion and persuasion. However, crucially, this persuasion will be situated within the broader context of an agalitarian public debate. Developing a risk dialogue is about the negotiation of difference. It is an attempt to reach consensus – or to accept divergence – out of a plurality of motivations, values and concerns.

The new wave of interest in methods of public consultation and involvement in Britain demonstrate great potential in facilitating this aim. By employing progressive methods of stakeholder involvement the objective of risk communications mutates from a one-way process of information dissemination, to the formation of social relationships of trust and respect [Otway 1992: 227]. Risk communication, viewed as a dialogical process amongst 'partners', creates the expectation of social relations characterised by mutual respect and reciprocity. In the short term, education and the provision of information are still set to figure as important goals of risk communication. However, if we conceive of risk communication as a dialogic process of social learning, we may transcend the traditional objective of education and information provision, and reach a terrain where multi-layered learning is possible:

> Risk analysts and managers may use risk communication as a means to learn from the public by listening to the concerns of local residents, public interest groups, and informed citizens. Public input is necessary to include risk-related properties other than magnitude and probability. Issues such as equity of risk bearing, catastrophic versus routine occurrence of losses, the circumstances of risk, and the ability of institutions to monitor and control hazardous facilities are excluded from formal risk analysis and hence not reflected in any risk calculation. Potential victims have a much better sense of risky situations and can communicate their concerns and observations to risk managers or regulators. Both information inputs – the scientific assessment of the probabilities as well as the public perception of the circumstances of the risk-bearing situation – are necessary elements of a rational decision-making process [Renn 1991: 458].

Having begun this paper by unpacking the sources of conflict which have produced a divide between expert and lay actors, I wish to end by re-affirming the pitfalls of imagining a neat boundary between two polarised parties. The assumption that experts and the lay public are compass poles apart in their interpretations of risk masks a more complex and messier social reality. In addition to the binaries of North and South, we need to explore the ground between and around these fields on the compass. A more sophisticated dialogue about risk which acknowledges and reflects the conditionality of risk knowledge is required. In matters of risk, we would do well to remember that the truth is not always 'out there'. Risk practitioners, scientists and politicians should not always expect – or be expected - to arrive at absolute and unquestionable conclusions about risk [Kajanne & Pirtilla-Backman 1999]. We must recognise that disputation and decision-making must take place in conditions of profound and open-ended uncertainty and that the meaning of risk is not intrinsic to a particular incident.

The development of meaningful political dialogue about risk must be sensitive to these issues of contingency and cultural difference. In short, there is no 'right' definition of risk. Rather, the meaning of risk will be infinitely contested, and reasonably so [Mythen *et al* 2000: 28]. The journey towards a more inclusive and interactive mode of risk communication will doubtless be marked by antagonism and discord. However, if the ultimate objectives of safety and security are to be advanced, this may be a trip worth making. Indeed, the institutional reflection which has surfaced in the aftermath of the BSE crisis perhaps suggests that we have already taken the first few steps.

References

[Adam 2000]   Adam, B: GM Foods and the Temporal Gaze. Sociology, 51: 125-142, 2000

[Beck 1992]   Beck, U: Risk Society: Towards a New Modernity. Sage, 1992

[Beck 1995}   Beck, U: Ecological Politics in an Age of Risk, Sage, 1995

[Bennett 1998] Bennett P: Communicating About Risks to Public Health: Pointers to Good Practice. HMSO, 1998

[Bennett 1999]   Bennett P: Understanding Responses to Risk: Some Basic Findings in Bennett P & Calman K. eds. Risk Communication and Public Health. Oxford University Press, 1999

[Breakwell, 2000]   Breakwell G: Risk Communication: Factors Affecting Impact. British Medical Bulletin, 56: 110-120, 1997

[Caplan 2000]   Caplan P: Risk Revisited. Pluto, 2000

[CESC Report 1997] Centre for the Study of Environmental Change Report. Uncertain World: Genetically Modified Organisms, Food and Public Attitudes in Britain. CESC, 1997

[Coles 1999]   Coles D: The Identification and Management of Risk: Opening up the Process' in Bennett P & Calman K. eds. Risk Communication and Public Health: 195-206, Oxford, 1999

[Covello 1983] Covello V.T: The Perception of Technological Risks: A Literature Review. Technological Forecasting and Social Change

[Cragg Ross Dawson Report 2000] Qualitative Research To Explore Public Attitudes to Food Safety. Cragg, Ross, Dawson Research, 2000

[Crook 1999]   Crook S: Ordering Risks in Lupton, D. ed. Risk and Sociocultural Theory: New Directions and Perspectives: 160-185, Cambridge University Press, 1999

[Dean 1999]   Dean M: Risk, Calculable and Incalculable' in Lupton D. ed. Risk and Sociocultural Theory: New Directions and Perspectives: 31-159 Cambridge University Press, 1999

[ESRC Report 1999] The Politics of GM Food: Risk, Science and Public Trust. Special Briefing No. 5. University of Sussex, 1999

[FAO/WHO Report 1998] The Application of Risk Communication to Food Standards and Safety Matters. Food and Nutrition Paper 70, 1998

[Fischhoff 1995] Fischhoff B: Risk Perception and Communication Unplugged: Twenty Years of Process. Risk Analysis 15: 137-145, 1995

[Fisk 1999]   Fisk D: Perception of Risk - Is the Public Probably Right? in Bennett P & Calman K eds. Risk Communication and Public Health: 133-140 Oxford University Press, 1999

[Flynn & Slovic 1995]   Flynn J & Slovic P: Expert and Public Evaluations of Technological Risks: Searching for Common Ground. Risk: Health, Safety and Environment 10: 333-358, 1995

[Frewer *et al* 1999]   Frewer L J., Howard C., Hedderley D. & Shepherd R: Reactions to information about genetic engineering: impact of source characteristics, perceived personal relevance, and persuasiveness. Public Understanding of Science, 8: 35-50, 1999

[Handmer 1995] Handmer J: Communicating Uncertainty: Perspectives and Themes in Norton T., Beer T & Dovers S. eds. Risk Uncertainty in Environmental Management. Australian Academy of Science, 1995

[Harris & O'Shaughnessy 1997]   Harris P & O'Shaughnessy N: BSE and Marketing Communication Myopia: Daisy and the Death of the Sacred Cow. Risk, Decision and Analysis, 2: 29-39, 1997

[Hinchcliffe 2000]   Hinchcliffe S: Living With Risk: The Unnatural Geography of Environmental Crises in Hinchcliffe S. & Woodward K. eds. The Natural and the Social: Uncertainty, Risk, Change: 117-153 Routledge, 2000

[ILGRA Report 1998]   Risk Communication: A Guide to Regulatory Practice. Health and Safety Executive, 1998

[Irwin 1995]   Irwin A: Citizen Science: A Study of People, Expertise and Sustainable Development, London, 1995

[Jungermann 1997]   Jungermann H: When you can't do it right: ethical dilemmas of informing people about risks. Risk, Decision & Policy, 2 (2): 131-145, 1997

[Kajanne & Pirtilla-Backman 1999]   Kajanne A & Pirtilla-Backman A. M:
Laypeople's viewpoints about the reasons for expert controversy regarding food
additives. Public Understanding of Science, 8: 303-315, 1999

[Kasperson, 1992]   Kasperson, R. E: The Social Amplification of Risk:
Progress in Developing an Integrative Framework in Krimsky S. & Golding D.
eds. Social Theories of Risk. Praeger, 1992

[Krimsky & Golding 1992]   Krimsky S. & Golding D: Social Theories of
Risk. Praeger, 1992

[Langford et al 1999]   Langford I., Marris C. & O'Riordan T: Public Reactions
to Risk: Social Structures, Images of Science and the Role of Trust in Bennett P &
Calman K. eds. Risk Communication and Public Health. Oxford University Press,
1999

[Leiss 1995]   Leiss W: Down and Dirty: The Use and Abuse of Public Trust in
Risk Communication. Risk Analysis, 15 (6): 685-692, 1995

[Lupton 1999}   Lupton D: Key Ideas: Risk. Routledge, 1999

[Macintyre et al 1998]   Macintyre S., Reilly J., Miller D & Eldridge J: Food
Choice, Food Scares and Health: The Role of the Media in Murcott, A. ed. The
Nation's Diet: the Social Science of Food Choice. Longman, 1998

[Mythen et al 2000]   Mythen G., Wales C., French S & Maule J: Risk
Communication and Risk Perception: A Critical Review. Manchester Business
School Working Papers No.411

[Otway 1992]   Otway H: Public Wisdom, Expert Fallibility: Toward a
Contextual Theory of Risk in Krimsky S & Golding D eds. Social Theories of
Risk. Praeger: 215-228, 1992

[Pidgeon, 2000]   Pidgeon N: Trust me, I'm a politician in The Times Higher
Education Supplement. September 15th, 2000

[Powell & Leiss 1997]   Powell D & Leiss W: Mad Cows and Mother's Milk:
The Perils of Poor Risk Communication. McGill-Queen's University Press, 1997

[Purdue 1995]   Purdue D: Whose Knowledge Counts? Experts, Counter-Experts
and the Lay public. The Ecologist, 25 (5): 170-172, 1995

[Ratzan, 1998]   Ratzan S: The Mad Cow Crisis: Health and the Public Good.
UCL Press, 1998

[Reilly, 1998]   Reilly J: Just Another Food Scare? Public Understandings and the BSE Crisis in Philo G. ed. Message Received: 128-145. Longman, 1998

[Renn 1998]   Renn O: The Role of Risk Communication and Public Dialogue for Improving Risk Management. Risk, Decision and Policy, 3 (1): 5-30, 1998

[Renn 1991]   Renn O: Risk Communication and the Social Amplification of Risk in Kasperson R & Stallen P. eds. Communicating Risks to the Public: 287-324. Kluwer Academic Publishers, 1991

[Scherer 1991]   Scherer C: Strategies for Communicating Risks to the Public. Food Technology: 110-116, 1991

[Slovic 1987]   Slovic P: Perception of Risk. Science: 236: 280-285, 1987

[Smith 1997]   Smith A. P: Consumer Information and BSE: Credibility and Edibility. Risk, Decision and Policy, 2: 41-51, 1997

[Tennant 1997]   Tennant D. R: The Communication of Risks and the Risks of Communication. Risk, Decision and Policy, 2 (2): 147-153, 1997

[Valenti & Wilkins 1995] Valenti J & Wilkins L: An Ethical Risk Communication Protocol for Science and Mass Communication. Public Understanding of Science, 4: 177-194, 1995

[Wales & Mythen 2002]   Wales C & Mythen G: Risky Discourses: The Politics of GM Foods. Environmental Politics. Spring 2002

[Weyman & Kelly 1999]   Weyman A. K & Kelly C: Risk Perception and Risk Communication. Health & Safety Executive Contract Research Report 248. HSE Books, 1999

[Wynne 1996]   Wynne B: May the Sheep Safely Graze? in Lash S., Sczerszynski B & Wynne B. eds. Risk, Environment and Modernity: 44-83. Sage, 1996

[Wynne 1996a]   Wynne B: Patronizing Joe Public. The Times Higher Educational Supplement. 12th April 1996

COMMUNICATION AND ELECTRONIC SAFETY

Safer Data: The use of data in the context of a railway control system

A. Faulkner CEng MBCS

CSE International Ltd, Glanford House, Bellwin Drive, Flixborough DN15 8SN, UK
Tel: (+44) 1724 862169
Email:agf@cse-euro.com

Abstract.

An increasing number of safety-related systems are configured to the application instance through the use of data. These systems typically use a static or slowly changing description of the infrastructure, in combination with a command schedule, instantaneous status data and a set of operational conditions. This paper uses the context of a railway control system to identify safety issues in the configuration of the control system and its reliance upon data from the external information systems.

Introduction

The safe operation of data-driven safety-related systems is likely to depend upon the correctness of the data. Nevertheless, data is often treated in a totally unstructured manner (often making error detection difficult) and very rarely is fault tolerance used to protect the system from data errors.

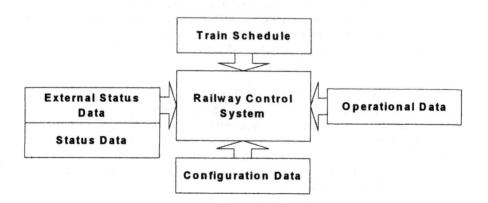

Figure 1: Data in the context of railway control system

The systematic production, development and maintenance of configuration data for safety-related systems receive little attention in both the literature and standards such as IEC 61508. In particular, many safety-related systems consist, wholly or partially, of generic software elements, which are adapted to a particular application by means of a description of the infrastructure, which describes the real world environment in which the system will operate.

The railway control system uses four categories of data (Figure 1). The author considers these categories, to be a generic description that could be used in other application domains. These four data categories are:

i)   *Configuration data*. A description of the infrastructure. This description is most conveniently regarded as "static" data in that it represents the entities from the real world, which change only in response to the action of maintenance or modification on these entities;

ii)   A *Command Schedule*. The schedule is used to describe the mesh of the required use of the infrastructure. For example a railway control system would use a train schedule to describe the planned movements of multiple trains across the rail infrastructure;

iii)   *Status Data* is provided through interfaces to external reporting systems and direct status information from connections to local sensors and other inputs; and

iv)   *Operational Data*. Individual operational conditions are commonly communicated to the railway control system via manual input. The operator receives these operational conditions through human communication interfaces such as telephone and fax. The set of operational conditions represents persistent restrictions on the use of the infrastructure through reports of flood, landslip or failed train.

A major weakness of the standards referenced in this paper is that they do not identify a requirement for the definition of a data model. These standards also fail to give guidance on appropriate forms of data representation or the mitigation of identified system hazards through the selection of high integrity data model elements.

In the railway control environment safety-related systems are arranged as an interconnected suite of computer-based applications. Traditional protection systems are employed to reduce the risk of train collisions by preventing multiple track occupancy; an example being the use of low-level systems known as *interlocking*. These combine railway infrastructure elements such as tracks, points, signals and train detection equipment, to allow railway signalmen to set routes and receive indications (instantaneous status data).

Interlocking systems provide a good example of the distinction between static and dynamic data. A *track circuit* is a form of train detection equipment, and unchanging information about a range of track circuits (their identifiers, their relationships to other track circuits, and their input/output addresses) is described by *static* configuration data. The current status of the track circuits (whether they are occupied or unoccupied, or whether they are serviceable or unserviceable) is described by dynamic data, and is obtained by sensor readings and other inputs.

Status information on the wider railway is provided through the train reporting systems external to the railway control system. These train reporting systems also provide access to data such as the train consist. The train consist is the description of the actual rolling stock undertaking an instance of a journey as described in the train schedule.

One of the functions of the interlocking system is to provide protection against multiple occupancy of individual sections of track. However, the interlocking system does not provide any protection where a train is 'out of gauge', for example, where a heavy goods train is to pass over a light railway bridge. Clearly, the railway control system needs to provide protection against these out of gauge trains to ensure the safe operation of the railway.

Product and application safety lifecycle

The railway control system will be subject to a safety lifecycle based within the national approval process and procedures. Railway focused functional safety standards such as CENELEC EN50128 identify the mechanisms of product development, but data-driven application development is poorly treated.

Product development based upon a generic 'V' lifecycle commences with requirements and an operational concept (how the product is intended to be used). Development planning establishes the process, procedures and approvals stage-gates required to implement the product. Following requirements analysis the system architecture is established, apportioning the system into components. This apportionment establishes how the system will achieve its requirements and usually not only describes the division between hardware and software, but also sets out the arrangement of hardware and software elements in sufficient detail for analysis of the proposed solution to take place.

The analysis of the proposed solution will require the identification of the safety requirements and the components of each of the safety functions of the system. The system architecture may be modified to take into account mitigations required to successfully implement safety functions to the satisfaction of the safety approvals authority.

The factory and site acceptance test plans will be created, together with the arrangements for verification and validation of the product. This paper observes that these standards identify product development requirements. These standards identify functional safety systems issues that can be adapted to application development.

Both the product and application developers may jointly undertake application development for the initial application of the product. Subsequent application of the product is likely to be undertaken by the application developers alone. Where the product is based upon Commercial-Off-The-Shelf (COTS) technologies the part played in the application development by the product developers may be minimal or even nonexistent. Many COTS computer systems are constructed from a generic software platform, which implements the necessary algorithms for the particular class of application, and is customised for a specific application. Common sense and good practice often dictates the use of configuration data in one form or another, rather than changes to the software code itself. Data-driven systems lend themselves to the deployment of multiple applications where the product component is tailored by data.

Implicit in the requirement for a data-driven system is a description of the data model and the data requirements of the system. Research conducted by the author has identified that data is often treated in a totally unstructured manner (often making error detection difficult) and very rarely is fault tolerance used to protect the system from data errors.

Data architecture referenced in standards and literature

The identification of system architecture is treated in the standards [IEC 61508, CENELEC 50128 and CENELEC 50129] but is based simply upon the need to provide a vehicle for analysis of the proposed solution. The identification of the system architecture should recognize the role-played by the configuration of the system components by data, as a majority of systems will contain these data elements to a greater or lesser extent.

European railway-specific standards such as references CENELEC 50126, CENELEC 50128 and CENELEC 50129 are, at the time of writing this paper, in the process of being issued. In particular CENELEC EN50128 section 17 "Systems configured by application data", proposes the following:

i)   That the development process should include a data lifecycle;

ii)   That the integrity of the tools employed should be appropriate to the Safety Integrity Level (SIL) of the system concerned; and

iii)   That the data lifecycle identifies a number of documents to be produced.

However the representation, in terms of the data model, and realisation, in terms of the production of the application data, and the subsequent maintenance of the data are not addressed.

Literature on the subject of data-driven safety-related systems is limited and does not address the issues of data architecture, data integrity, of data definition.

Welbourne and Bester [Welbourne 1995] identify several categories of data:

i)   Calibration data, for example alarm and trip levels;

ii)   Configuration data, for example display screens for a specified area; and

iii)   Functionality data, for example control states to be taken up at failure.

The Welbourne and Bester classification is not useful in expressing the role of data in the safety-related system.

Data within a hierarchy of interrelated systems

From the descriptions given in the introductory paragraphs the railway control system is a component within a systems context. The railway control system has peers, subordinate and supervisory systems. Data is shared amongst these systems. Configuration data is used in all of these railway systems to represent the same physical infrastructure. Each of these systems within the systems hierarchy requires data to perform a range of systems functions at differing levels of abstraction and detail. The interlocking system uses physical data concerned with the attributes and characteristics of installed infrastructure (Physical model). The railway control systems uses similar configuration data but is more concerned with the arrangement of these physical equipments to describe routes across the infrastructure (Operational model). Where the layout of the railway does not permit the unique identification of a single route between an entry and an exit signal some form of route preference weighting (data) is required. This route preference weighting may not be a single value, but be an algorithm that takes into account type of train, class of service and final destination. These additional abstractions are used to infer the relative priority of one route over another (Planning model).

Information used for high level planning may only require broad data that describes the overall characteristics of the railway infrastructure between two geographical locations. This information may be summarised as the number of train paths per hour available for an identified class of service. This abstract information may form the basis strategic decision making with a time horizon of 6 to 24 months. Strategic information is used in the context of corporate management. Information volumes at the corporate level are generally small.

Where operational conditions persist for a period of time, tactical decisions are required. These persistent operational conditions may be as a result of climatic conditions such as flood or landslip of an embankment. Information at this level of the system is generally tactical and tries to deliver the short to medium term goals of the organisation. The business unit uses this tactical information in the delivery of the train service. In the event of a major incident or accident the business unit will be come directly involved in the operation of the railway.

Each business unit may operate a number of railway control systems. These systems typically use data with a time horizon of –48 hours to +72 hours. The train schedule (timetable) information is typically applicable to a 6 month period (Winter or Summer) with regular updates. Short-Term Planning (STP) and Very Short-Term Planning (VSTP) changes to the train schedule are communicated to the train control system. The purpose of the model described in the following paragraphs is to articulate the differing data requirements of each of the levels of the railway organisation. Each of these layers may use and re-use data entities in other layers. In the context of functional safety, changes to individual data entities may not result directly in an accident, but in the wider context may result in some loss.

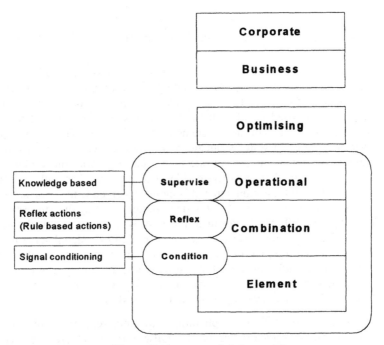

Figure 2: A proposed data model

This model identifies a number of layers within the data hierarchy and infers a systems hierarchy based upon the role and nature of the system.

The 'Element' layer represents single instances of elements of rail infrastructure, which are considered as signals, point machines, train detection equipment and track.

The 'Combination' layers represents the use of these infrastructure elements to describe a concept such as a route. A route would be created from an entry signal, one or more track sections, and may include one or more points machines and an exit signal. The route would also contain train detection equipment. The combination layer may also undertake some signal conditioning processing such as the translation of a 4-20 ma sensor into a meaningful value.

The *'Operational'* layer represents the use of the combinational layer elements to effect control over the rail infrastructure to facilitate the passage of trains. Where a protection system does not require the intervention of a operator these protection systems are described as reflex actions and may exist at the border of the combination / operational layers. These reflex actions are essentially rule based.

The operational layer requires knowledge-based actions, in the context of a conventional railway control system this layer would represent the signal box, with the signalman and a display panel or workstation.

This operational layer should be the highest layer at which a safety function should be implemented. This boundary depicted in figure 2 by the box surrounding the element, combinational and operational layers.

The *'Optimising'* layer is used to represent the use of an operational sequence to achieve the efficient utilisation of the rail infrastructure. An example of an optimising condition may be where a supervisory function (Operator) may take advantage of the railway system to regulate trains to recover time for a delayed train. This layer may be automated as in the case of Automatic Route Setting (ARS) equipment which implements a train schedule (timetable) to ease the workload of the signalman. The optimising layer needs to take into account knowledge of current and some future operational conditions. These operational conditions may be restrictions on the use of the rail infrastructure due to planned or unplanned maintenance.

The *'Business'* layer represents the tactical implementation of the longer-term strategic goals of the organisation. These tactical needs may be required due to persistent operational conditions such as flood, landslip or major accident.

The *'Corporate'* layer represents the strategic planning of the use of the rail infrastructure. This strategic layer has a time horizon of greater than 6 to 24 months.

The data model is used to illustrate the nature of data within the railway control organisation.

Data faults

The data model in Figure 2 should be represented in a data description, which facilitates analysis. Good systems development process aims to reduce the faults within the final product. In systems requiring a high integrity additional measures must be taken to overcome the effects of faults. Broadly these may be divided into four groups of techniques:

i)   Fault avoidance;
ii)   Fault removal;
iii)   Fault detection; and
iv)   Fault tolerance.

Fault avoidance aims to prevent faults from entering the system during the design stage, and is a goal of the entire development process. Fault removal attempts to find and remove faults before a system enters service. Fault detection techniques are used to detect faults during operation so that their effects may be minimised. Fault tolerance aims to allow a system to operate correctly in the presence of faults.

Safety-related systems are frequently constructed from a generic hardware and software platform, which is adapted to a particular use by providing it with data. This data is an essential part of the design and to ensure that the system as a whole has the correct behaviour and safety integrity the data must be developed and verified with as much care as the software and hardware. The railway control system uses four categories of data. Faults within each category of data are discussed in the following tables.

Harrison and Pierce [Harrison 2000] identify the classifications of faults in the data used to describe the railway static infrastructure (as shown in Table 1).

Fault	Description
Omission	An infrastructure entity is not present in the control system data set
Spurious data	A non-existent entity is present in the data set, this may also include duplicated entities
Positioning faults	For example an entity is represented and addressed correctly, but its physical position is incorrect
Topological faults	All entities are present, but they are connected in a way which may be plausible, but incorrect
Addressing faults	An entity is correctly located and labelled but is connected to the wrong field devices
Type faults	An entity is connected and labelled correctly but is recorded with an incorrect type identifier
Labelling faults	An entity is located and addressed correctly, but is assigned the wrong label in the data model
Value faults	A scalar attribute of some entity in the configuration data has the wrong value

Table 1: Data faults in infrastructure data

Status data is derived from two sources. Equipment connected to the control of the railway control system and external train reporting systems. The equipment directly connected to the railway control system are infrastructure monitoring devices and the interlocking(s).

External data systems should not be assumed to use a single representation of the railway in a compatible set of data models. Indeed many of these external systems have been created over a substantial time period. Data passing through these external systems will be subject to a number of adaptations and transformations. Train reporting systems will record the progress of the train. This train progress data is based on known timing points at specified locations on the rail infrastructure. These timing points represent the coincidence between the train schedule description (Planning model), the physical arrangement of the railway and the actual route taken by the train (Operational and Physical models).

The data faults classified below (in Table 2) refer to the status data obtained directly from the equipment connected to the railway control system. In this case this would be the interlockings, infrastructure status equipment such as rock-fall detectors, and infrastructure equipment such as 'hot-axle-box' detectors (HABD) or 'wheel impact load detection (WILD) equipment.

Fault	Description
Mode faults	The operational mode for one or more component is incompatible with the current operational mode
Sequential faults	Data presented to the railway control system is '*out of sequence*'
Combinational faults	Data presented to the railway control system is incomplete to meet the requirements of a predetermined condition
Propagation faults	Data transmitted to the railway control system is changed or corrupted on receipt
Timing faults	Data is presented to the railway control system earlier or later than expected by the system
Volume faults	The railway control system is flooded with or starved with data

Table 2: Data faults in status data

In addition external systems may provide data fault types (listed in Table 3) [Faulkner 2001]

Fault	Description
Existence	A data reference provided by one external information system cannot be fulfilled by another information system
Reference faults	The wrong data reference is provided resolving information which does not represent the required train
Availability	An external information system is not available (off-line) at the time the information is requested
Inconsistent	Data requested from more than one external information systems is inconsistent – which data will be used?
Timely	Data is not supplied until after the event

Table 3: Data faults in status data provided by external systems

Operational data is passed to the railway control through the use of human interaction via telephone or fax. This operational data represents persistent restrictions on the use of the infrastructure. This data will be used to place tokens, typically on infrastructure elements within the railway control system to act as semi-permanent bars or reminder devices to operator.

A work site on the infrastructure requires restrictions to trains entering that section of infrastructure. Where the rail is being replaced, only specialised engineering vehicles should be allowed into the section. The controls on entry to such work sites are managed (in-part) through the use of these tokens in conjunction with operating process and procedures. Operational data faults classified below (in Table 4).

Fault	Description
Existence	A data reference provided cannot be matched to an element(s) of infrastructure
Reference faults	The wrong data reference is provided placing a restriction on the wrong item of infrastructure
Type faults	The type of restriction is incorrect
Inconsistent	Data used to describe the restriction is inconsistent with the description used of the infrastructure
Timely	A restriction on the use of the infrastructure is applied or removed at an inappropriate time
Mode faults	The operational mode for one or more component is incompatible with the current operational mode
Sequential faults	Data presented to the railway control system is *'out of sequence'*
Combinational faults	Data presented to the railway control system is incomplete to meet the requirements of a predetermined condition
Propagation faults	Data transmitted to the railway control system is changed or corrupted on receipt

Table 4: Data faults in operational data

In addition external systems may provide train schedule data with the fault types (listed in Table 5).

Fault	Description
Existence	A data reference within the train schedule does not exist
Reference faults	The wrong data reference is provided resolving information which does not represent the required train journey
Availability	A journey described within the train schedule is unavailable to that class of train or type of service described in the train schedule
Inconsistent	Data used to describe the same train type of type of service is described in a number of inconsistent forms
Timing	Train schedule data is not supplied until after the train has commenced its journey
Propagation faults	Data transmitted to the railway control system is changed or adapted before presentation.

Table 5: Data faults in the train schedule data

System boundary issues

Data used within a hierarchy of interrelated systems will reference infrastructure elements and make use of the infrastructure elements in differing ways. A train schedule will identify the start location, time and date, the intermediate points and times and the destination and arrival time. To describe the journey the train schedule does not need to describe the complete journey. The execution of the journey will be influenced by the operational status of the railway. To satisfy the journey description expressed in the train schedule the journey may take a number of compatible routes.

The train schedule describes a logical journey, which respects the topological connection description of the railway (Planning model). The execution of the journey uses a physical description of the railway in the context of the current operational status of the railway and the train undertaking the journey (Operational model). These two representations of the railway are both valid and represent some of the translations and adaptations which the data within the railway control system.

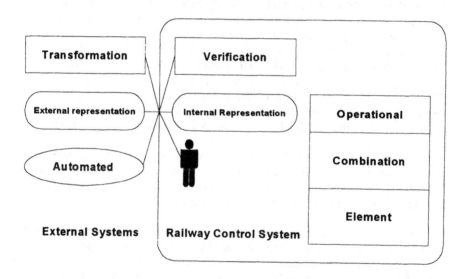

Figure 3: System boundary issues

Data presented at the boundary of the railway control system by an external system will be transformed or adapted from the external representation to the internal representation of the railway control system. This transformation or adaptation may occur in some automated function of may require manual intervention. This data will require verification. Analysis of the railway control system design is required to establish the consequence of faults in this data as this data passes across the boundary of the railway control system. Further analysis should also establish the sensitivity to changes in this data. Such analysis will facilitate the definition of 'properties' or rules by which faults in the data may be detected at the boundary of the system.

Discussion

This paper has identified four categories of data used by a railway control system, to describe the system and its use of data (Figure 1). The extent to which this system places reliance upon these categories of data can only be established through analysis. This paper has provided a number of classifications of data faults for each category of data to aid in this analysis (Tables 1 to 5).

This paper has developed the information model (Figure 2), based upon the use of data by a railway organisation. The information model is used to provide a framework in which to express the use of data representing the rail infrastructure by information systems at different levels of abstraction. The information model proposes an upper boundary for data that is to be used by safety-related functions. This boundary is proposed to ensure that changes to data at a higher level in the information model should not adversely affect the functional safety of the system.

The railway control system does not contain a single unified model but a number of models that are associated with specific groups of system functionality. (a Planning model, an Operations model and a Physical model). These models are arbitrarily coincident where these models overlap. The physical and the operational model share common nodes where train paths divide and join at junctions.

Viewing the data model in the context of good systems practice infers *information hiding*. Access to individual data attributes or elements should be via a predefined interface. Data within each layer should contain 'properties'. These properties represent rules, which are used to aid verification.

Direct observation of disruption to the rail service during the introduction of the seasonal timetable (Winter/Summer) identifies that during this period of change operational data is being rediscovered. This operational data is a result of the imperfections of the data models and representations used within the information situation. Perturbations to the train service at such times are based in the resultant modifications to both train schedule and operational data.

Verification of data remains a primary issue. Data quality, completeness and consistency at the railway control system boundary are likely to be beyond the influence of the implementer. The data boundary model (Figure 3) identifies the requirement to manage data across the system boundary. External systems, in the context of the railway control system, may provide data that is inconsistent, incomplete or plain incompatible with the intended use by the railway control system.

Data within the railway control system may be safety-related as it contains information, which adjust the likelihood (opportunity) of an accident occurring. In the case of the train schedule, implicit in the arrangement of data will be a sequence of the train service. Each train service will be associated with a train consist. Each instance of the train journey will have attributes based upon the layout of the infrastructure and capability of the train. Train schedule sequencing will influence the efficiency of the operation of a junction. Perturbations to the railway will cause disruption to train running that may require the train schedule to be re-planned using a different type of scheduling system (based upon train priority, arrival time, or train sequence).

Conclusion

The safe operation of data-driven safety-related systems is likely to depend upon the correctness of their data. However, tools and techniques, which would assist in the creation of data models for such systems, are not treated in the literature, or in standards such as IEC 61508.

This paper has illustrated some of the issues related to the use of data, through a brief example - a railway control system. Through the context of the railway control system this paper has identified a number of data faults. This paper has observes that safety-related systems often contain a number of data models. This paper has also presented a number of frameworks to assist in the analysis of data driven systems to determine the integrity requirements of data used by the safety-related system.

The production of any safety-related system should include the development of a system architecture that contains not only hardware and software elements, but also components representing its data. All these elements should be analysable and should lend themselves to verification and validation.

Although a number of data fault modes have been identified, further work is required to provide guidance on tools and techniques that are suitable for the production of high integrity data-driven systems.

References

CENELEC EN50126 Railway Applications – The specification and demonstration of dependability – reliability, availability, maintainability and safety (RAMS). Comite European de Normalisation Electrotechnique, Brussels, 1999

CENELEC ENV50129 Railway Applications – Safety-related systems for signalling. Comite European de Normalisation Electrotechnique, Brussels. May 1998.

CENELEC EN50128 Railway Applications – Software for railway control and protection systems. Comite European de Normalisation Electrotechnique, Brussels March 2001.

[Harrison 2000] A. Harrison and R. H. Pierce: Data Management Safety Requirements Derivation. Railtrack: West Coast Route Modernisation Internal report. June 2000. Railtrack PLC, 2000.

IEC 61508 Functional Safety of electrical / electronic / programmable electronic safety-related systems – Part 1:2000 General Requirements. Geneva: International Electrotechnical Commission, 2000.

[Welbourne 1995] D. Welbourne and N. P. Bester: Data for Software Systems important to safety. GEC Journal of Research, Vol. 12, No. 1, 1995.

e-Technology
Excitement, Enlightenment and Risk

A Hutson-Smith
Data Systems and Solutions
www.ds-s.com

Abstract

This paper considers the use of e-technology in the markets where safety is of primary consideration. The paper considers the attitudes of both the developer and sponsor of such systems, the benefits realised and the risk introduced by their use.

For ease of writing I refer throughout the paper to three terms:

- **Safety Significant Industry** - any area of industry where safety is of primary importance. Obvious examples would be rail, aerospace, nuclear and process industries.
- **e-Business** - the integration of systems, processes, organisations, value chains and entire markets using Internet-based and related technologies and concepts.
- **e-Technology** - any technology employed in e-Business systems. Typically Commercial-Off-The-Shelf (COTS) based products, for instance web browser technology.

1 e-Technology within Safety Significant Industries

Today, in order to execute their day to day business the majority of companies word-wide have PC networks, e-mail and marketing web sites. e-Business systems are rapidly becoming an essential route to market for many businesses. For instance, the majority of the developed world's retail and banking is now supported by e-business systems and in many instances dominated by it. Industry and manufacturing too are rapidly adopting e-business systems to streamline procurement and supply chain processes. There is much excitement around the opportunities introduced by web enabling traditional systems. It is however notable that the adoption of e-business and e-technology within the Safety Significant Industries is relatively slow and in some instances non-existent.

In order that we may assess why the adoption of e-technology in Safety Significant Industries is so low, two things must be considered; firstly, the factors that differentiate Safety Significant Industries from other areas of industry and secondly, the nature of e-technology and its possible marriage with the traditional technologies employed in this market.

The following characteristics are extremely pertinent to Safety Significant Industries:

- Data integrity and tracking of history for certification
- Highly invested processes and procedures
- High technical complexity
- Need for detailed domain knowledge
- Low volume with significant configuration options
- Regulation and certification requirements that are significant and in some instances have a bias towards a traditional development lifecycle
- Conservative with respect to business processes
- Significant security and data access issues – often restricted connectivity
- Large organisations often operating a restricted preferred supplier system

Clearly the barriers faced by a software system developer wishing to enter any part of the Safety Significant Industry are high. The ability for a typical e-Business developer to enter the market is very limited.

Typically e-technology solutions have the following characteristics:

- High reliance on integration of Commercial off the Shelf (COTS) software systems
- Targeted at high volume user markets
- Solutions that tend towards modifying a customers process rather than the software product
- A Focus on time to market rather than absolute integrity of software
- Software upgrades and evolution which is fast and furious
- System shelf-life that is relatively low
- Recent 'dot-com' events have shown company and staff shelf life are potentially equally as low
- Requirements that come from a top down business capture process rather than from a detailed technical specification
- Development time-scales that are very tight and systems which are usually developed using rapid application techniques such as Dynamic Systems Development Method (DSDM)
- Functional testing which is usually performed at a system level with positive confirmation of presence of desirable operation, rather than demonstration of absence of undesirable operation
- Minimal destructive or abnormal scenario testing
- Issues such as security are often addressed late in programme in a 'penetrate and patch' fashion

Clearly the typical characteristics of e-technology systems are largely unsuitable for use in the Safety Significant Industry. In order that a structured argument for the use of an e-technology-based system could be constructed the development lifecycle, processes, use of pre-defined components and tools would need to be rigorously assessed.

A recent Department of Trade and Industry (DTI) study [DTI 2000] considered the impact of e-business on the UK Aerospace market and noted that much of the apparent 'e-laggard' attitude was due to the inability of typical solution providers to enter the market.

It should however be noted that the large players in the US anticipate that much of their business both procurement and operationally will be significantly reliant on e-technology. In some areas such as supply chain systems it may well become a prerequisite that a supplier is 'e-enabled'. The reliance of the UK industry on the US is without doubt. The large-scale adoption of e-technology in the Safety Significant Industry is inevitable.

Despite the difficulties presented by the nature of the Safety Significant Industry and the level of rigour it demands, the opportunity to introduce and exploit e-technology remains high. The introduction of e-technology systems present the ability for an organisation or group of collaborating organisations to integrate disparate systems, realise great advances on presentation and delivery of data and reduce the potential for duplication and even human error. It would, however, be difficult to ever imagine an argument being approved for the use of an e-technology system to remotely configure a Full Authority Digital Engine Controller (FADEC) via the web. The realistic opportunity for the introduction of e-technology in the Safety Significant Industry is perhaps as follows:

- **Information Management and Presentation** (activities associated with knowledge management, research and development and marketing)
- **Commerce** (sales and procurement – supply chain management)
- **Project Support** (development project collaboration – sharing planning and engineering design data)
- **Operation Support** (configuration management, asset management, operation control, maintenance, repair and overhaul activities)

So is there a risk to the safe operation of a company or its products if it introduces e-technology systems? Obviously a marketing presence on the web introduces no risk to the safe operation of any organisation, the risk comes where an e-technology system is employed in a fashion where reliance on its integrity and the data it manipulates is assumed, potentially without foundation or argument.

2 System Context Creep

In considering the use of e-technology in Safety Significant Industries, I have not only spoken to the developers of these systems, but also to the end users of the systems. The level of adoption of e-technology across Safety Significant Industries varies but one significant issue seems common – that of 'context creep'.

We are all perhaps familiar with requirement or scope creep, where the functions of a system grow as its development progresses. Scope creep is common; it is not peculiar to any market, methodology or programming language. Indeed, scope creep is even experienced in the tightly bound development of typical safety significant systems, such as surveillance and control systems, which in contrast to business systems have relatively clear requirement definitions from the start. For example, a safety critical weapon controller may take on additional functionality: new missile release responsibility, an increased number of critical parameters to be measured, or perhaps a whole new data logging facility. It would, however, be rare for additional functionality to be introduced that would raise the Safety Integrity Level (SIL) requirements of a typical safety significant system (I personally know of no instances where this has happened). All other issues aside, the important fact is that the application of rigorous development techniques are designed to ensure that the introduction of new functionality is visible, appropriately analysed, implemented and tested. A rigorous safety critical system development process should ensure that no potentially hazardous functionality is introduced either by stealth or ignorance.

Context creep is a more subtle phenomenon where the reliance on a system and its data changes over time, potentially without change to the system itself. This phenomenon has been seen for years with management information systems, used to support operations such as asset tracking. It is now an even more acute problem as large stand-alone systems are integrated into, or replaced by, e-technology based systems. By their very nature web based systems bring together huge data sets, and potentially deliver the information to a vast audience. Simply put; they are useful, and likely to be employed by an ever-increasing audience, who seek to realise the value-added potential of the newly available data.

2.1   Illustration of Context Creep – A Factory Operation Support System

This section considers how context creep may occur with an e-technology system. A fictional illustration is used; however, the individual scenarios, actions of the personnel, and resulting problems are based on programmes I have had direct visibility of or have studied [Glass 1997][Young1999].

The illustration (depicted in Figure 1) assumes a large factory with a number of safety significant processes.

Three distinct points in time are considered to illustrate the shifting context of the e-technology system and in-turn the safety significant role it eventually plays.

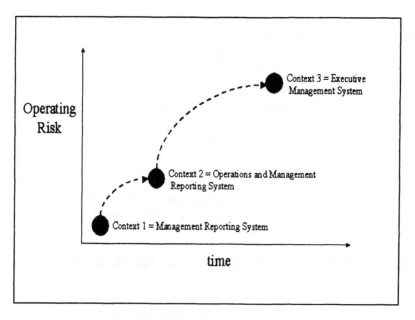

Figure 1 – Illustration of Context Creep

Context 1: Role = Management Reporting System

As mentioned above our example factory has a number of safety significant processes. These processes are supported in part by complex electromechanical devices. The devices operate in a harsh factory environment and are subject to wear and potentially total failure through fatigue. Failure of these devices in combination could be catastrophic. The devices have been designed to meet the specific need of the operating company. A number of manufacturers provide the devices; each uses a slightly different solution although some components are common. The operating company has very regular inspection and servicing schedules on the devices, replacing possible wear components well within the manufacturers calculated Mean Time Between Failure (MTBF). Practically no historical failure rate data is available for the components or composite devices. The factory maintenance staff currently removes used components as appropriate, but continues a simulated operational analysis in a test bench environment. Data is logged both from the various process lines and the test environment.

The division manager responsible for the factory has tasked his IT group with producing an e-technology based Management Reporting System (MRS) to serve two principal objectives:

- Provide a web based reporting system used to identify possible anomalies with components such as unusually high or low life.
- To allow the factory's purchasing department to select the supplier of electromechanical devices that offers best long-term value.

A huge volume of data is collected centrally by the Management Reporting System. Access to the system is provided to all engineers, manufacturing staff and the operator's maintenance body. All parties can run configurable queries over the web to produce highly visual reports that present the specific set of data that they are interested in.

The web-based MRS has been designed on a budget and employs a high level of COTS software packages (in line with corporate policy), although unusually in this market some 40% of the system is still bespoke. The coupling between COTS and bespoke packages is high. As the MRS is intended to be used only in the short-term the bespoke software is largely undocumented. The reports and resulting decisions have no bearing on safety and are considered to be only of commercial importance. Needless to say the managing director of the operating company is very excited by the graphs he can plot from his desk at leisure, but the managing director of each of the suppliers is less enthusiastic!

Context 2: Role = Operations and Management Reporting System

The system is now well respected. It is occasionally unavailable due to 'server problems' but the data has never been corrupt and the system is very useful in selecting the best electromechanical device supplier. It is considered by all to be more user-friendly than the paper-based system that the safety people mandate. About nine months into the use of the MRS two significant points were raised by the operator's maintenance team. Firstly, it was found that a number of critical components seemed to fail significantly earlier if they had been subject to high temperatures; secondly, if operated at ambient temperatures but under high pressure conditions the component life is significantly extended. It is considered by factory management that these findings indicate that there is potential for increased safety through small-scale preventative maintenance of the process lines operating in high temperature environments, and more significantly, extend routine maintenance intervals of process lines operating in high pressure environments. It is calculated that if acted upon these findings could significantly reduce factory down time. Corporate funding is obtained to 'beef the system up' to ensure its accuracy and reliability. The Management Reporting System (MRS) is re-branded the Operations and Management Reporting System (OMRS).

The head of software within the operating company has a healthy suspicion of software. The COTS based software auto-query tool utilised by OMRS is no exception (the only other instance where the software tool supplier could claim any significant use was with marketing and insurance companies). It was decided to carry out a manual analysis on the data captured directly from the electro-mechanical devices on a sample basis. No anomalies were detected between the manually calculated figures and those presented by the OMRS (although the data and findings were not recorded under any quality management system). Given the apparent absence of errors it was not considered necessary to 'beef the system up' except by adding more processor power and splitting the system into application, database and web servers – pretty standard stuff. The rest of the corporate funding is spent on 21" high-resolution liquid crystal monitors to allow the management team to view the graphs from the OMRS which the latest upgrade of the COTS auto query tool can now draw three dimensionally!

Data reporting from the trusted OMRS is now used across the company in all factories to support decision making on maintenance cycles and scheduling. No executive authority is given to OMRS as it is cross-checked manually by highly skilled personnel.

Context 3: Role = Executive Management System

Some time later, with great surprise to the company, a large number of the highly skilled personnel have left and the new recruits find very few anomalies with the OMRS which is now reported to have 'settled down'. The managing director that commissioned the original system has now employed management consultants to look at his business and suggest any possible streamlining. Following months of work they suggest two principle measures. Firstly, remove the costly software department head (after all his aerospace background is hardly relevant here!) and, secondly, fully automate the labour intensive calculation of maintenance cycles and scheduling.

The managing director is congratulated for his innovation and excellent use of corporate funding. He is proud that in the last two years his division has only had to hand craft fifty new lines of code, all done by a junior in JavaScript, whilst the rest is in COTS which is in line with corporate policy. The OMRS is grandly re-branded the Executive Management System (EMS).

Unfortunately, the newly established industry regulator (himself of aerospace background) is deeply concerned by the lack of tangible evidence within the company that the Executive Management System is suitably qualified to make unchecked decisions that could lead to catastrophic failure of the factories safety significant processes. This, coupled with the lack of any mention of the EMS or any of its other incarnations in the company's Safety Management System, causes him to place a closure order on the factory process line until a suitable manual system can be established using 'suitably skilled personnel'.

The management consultants are called in again to help. They recommend that they employ a software safety consultancy company to assess their position and establish a way forward. They recommend a small company headed by an individual that has both extremely relevant factory domain knowledge and highly valuable experience in other markets where software safety is of primary consideration.

2.2 Conclusion

As mentioned previously the illustration is fictional but the thread of events is, perhaps, not far-fetched. The following key points should be noted:

- e-Technology systems can readily be interfaced to a host of very flexible COTS software. In the illustration this is a flexible visualisation and reporting tool. It is inevitable that tools providing both significant efficiency gains and excellent visualisation are likely to become a primary source of information, and, in many instances, without proper control will replace proven manual sources of information.

- A side effect of automated analysis and presentation of data is the shift in the use of skilled personnel. The transition from an executive role to one of monitoring (confirming the correct operation of a computer system) is likely to significantly demotivate personnel and is often the root cause for loss of personnel and important domain knowledge.

- The sponsorship of e-technology systems is typically from high level, often at corporate level. Essentially, the demand for the system is from a business rather than engineering source. As a result, the perception is that everyone's reputation depends on the speedy delivery of the system. The result can be the development of these systems being given priority over other programmes; developers are afforded waivers on process, and all too often timescales set are insufficient to allow the application of rigorous techniques.

- As a web-based system allows delivery of data to a broad audience, it is essential that the use and reliance on this data is tracked and proactively managed.

3 Guidelines and Standards

Other than for marketing and supply chain, the use of e-technology in the Safety Significant Industry is not yet commonplace. The e-technology market is relatively immature and remains practically self-regulating. e-Technology guidelines and standards drawn-up to date address issues such as usability, accessibility and data interchange, and extend to address more relevant issues such as security and data integrity. There is little available to help the more specialised needs of the developer of a potentially safety significant e-technology system.

In contrast techniques applied to safety significant software have been maturing for many years. In the majority of instances safety significant software is appropriately developed, tested and maintained. Regulatory bodies are well established and intelligent both with respect to appropriate application standards and knowledge of the domain in which the system will be operated. The relationship between the regulator and the developer is becoming ever more collaborative, with large-scale developers operating well documented safety management systems.

In recent years the commercial and operational arguments of using commercially available pre-developed components in the Safety Significant Industry is becoming compelling. Clearly the use of these pre-developed components in a Safety Significant Industry will require appropriate assessments and well-structured arguments. A great deal of work is being done to draw-up guidelines and standards for addressing issues introduced by the use of pre-developed software (e.g. COTS and Software of Uncertain Pedigree (SOUP)). An excellent example is the work done by Adellard on behalf of the HSE [Bishop, Bloomfield, Froome 2001][Jones, Bloomfield, Froome, Bishop 2001]. It is perhaps this area of research that provides the best foundation on which to construct appropriate guidelines and procedures for development, test and maintenance of e-technology systems [Voas 1998][Voas, Charron, Miller 1996].

4 Possible Measures

It is difficult to predict how the functions delivered by an e-technology system will evolve over time. It is perhaps even more difficult to predict what reliance will be placed on the data it generates. At project launch, even where requirements or context creep is considered likely, it would be difficult to justify the application of the suite of measures mandated by safety significant standards. In many instances the techniques mandated could not be applied due to use of the proprietary software products, the use of languages such as Java and the lack of appropriate analysis support tools.

A safety significant software system is licensed for use only in the presence of extensive evidence that the developer has taken appropriate steps to ensure the integrity of the system, or can draw upon external evidence of reliability gathered through extensive use in the market place.

Where in time an e-technology system may require a safety assessment or be referenced as part of a safety management system then our aim should be clear:

> *To provide an extensible framework for extracting evidence of what rigour has been applied in developing the system, providing a suitable foundation for constructing a component level safety argument, and, as necessary, adding additional engineering rigour.*

Taking the illustration in section 2 of the Executive Management System, the company has found itself in a position where a safety argument is required. As you will recall, the system is a combination of tightly coupled COTS and bespoke development; the bespoke development is undocumented. In the absence of the ability to perform a detailed code assessment in a 'white box' style the company are left only with the option to treat the system as a 'black box'. However, this approach would as a minimum need a detailed requirement definition and, ideally, evidence drawn from a large user base of the system having delivered reliable operation. Neither are available. In practical terms they would be left with little option but to develop a system from scratch.

Techniques applied to safety critical development, scaled as appropriate, can represent real benefits in any development with relatively little cost. An illustration of this is the use of requirement management tools such as DOORS. We have found that use of requirement management tools even on a web-based marketing portal has been beneficial both in terms of work allocation, control and managing maintenance issues such as impact analysis. In seeking to build a suitable basis for any e-technology system that may in time have a safety significant role I believe as a minimum the following should be considered:

- **Requirements and Design** – e-technology developments typically employ a highly iterative lifecycle. In many instances production systems are developed as rapid and often undocumented output from stakeholder workshops. Whilst this may lead to a satisfactory outcome in the short term, in the longer term it is difficult to trace the source of requirements and any basis of decisions. For larger more complex systems this approach becomes unmanageable through development and, more significantly, can result in an almost unmaintainable system. If the requirements to code, and the internal partitioning of code components (either bespoke or COTS) cannot be demonstrated then there would be little basis on which to construct a safety argument for the system as a whole, or any component of the system. Our experience has shown that the use of a requirement tool with a case tool supporting an appropriate design language such as Unified Modelling Language (UML) can be deployed at very little cost. The ability to identify components of software in terms of their use in the system and to identify their boundary and relationship with other components is obviously essential for performing any detailed assessments such as software hazard analysis. Characteristics such as vertical architecture partitioning, loose coupling and division of COTS services from bespoke code will all aid efficient goal-based assessment, and thus construction of safety arguments around tightly bound sets of software components.

- **Evidence** – The ability to show evidence of reliability over a wide user base may be key to constructing an acceptable argument for the use of an e-technology system in a safety significant role. The use of a formal Fault Recording and Correction Action System (FRACAS) need not be a large overhead. Ideally it would just be part of the overall configuration management system. User community reporting is also key, as is the recording of the user's profile with respect to the system functionality he is known to exercise.

- **Test and Assessment** - It is neither always necessary nor economical to test all parts of an e-Technology system. However the developer of a system that is possibly going to influence safety should at least set out a clear agenda for what will be tested, how, and what the merits of this testing will be. We have used a test appraisal technique where discrete elements of functionality are assessed at the time of design to establish the most efficient technique for demonstrating the components' correct operation, and possibly the risk introduced by not applying certain techniques. A basic matrix system is used where the decisions are recorded. Again, if a requirement management tool is used then this data may be linked to requirements, design and code components. Also, any change of the system's context may be easily assessed and the need for additional tests kept to a minimum.

- **Safety Documentation and Standards** - The most basic of project quality systems will include a quality plan. The development of any e-technology system in a safety critical industry would benefit from a section that discusses safety. Even if the system, in its initial context, has no safety significance the section will remain and each review of the quality plan will force a review of the situation.

  If at any point it becomes likely that the e-technology system may play a part in safety then a candidate standard should be nominated. Whilst there is currently no standard wholly suitable for the development of an e-technology system, seeking partial compliance with a standard and constructing a mitigation argument for not meeting areas of the standard is more likely to gain approval that an argument with no reference point.

- **Quality Management** – There are a number of relatively low cost audit and static analysis tools. These can be easily administered to provide a useful and continual mechanism for identifying instances of significant deviations from defined criteria or the appearance of undesirable characteristics. The use of highly visual tools such as those providing a Kiviat diagram output (see figure 2) allow for very cost effective 'assessment at a glance' – a very powerful tool when balancing budgets with risk of potential re-work should the system context change in the future.

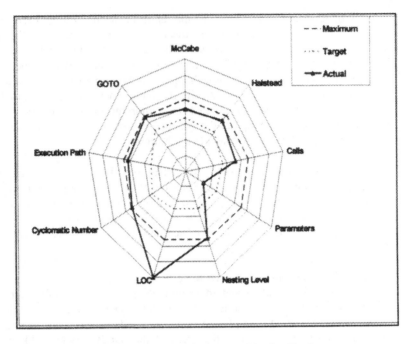

Figure 2 – Software Component Metrics Graph

5   Conclusion

Without question the benefits of e-technology systems are significant. The ability for any company to collaborate with other parts of its organisation or potentially external partners and even customers is significant. e-Technology offers the potential to compile data from diverse sources, add value to the data and disseminate it consistently and efficiently, and presents companies with new opportunities for improving almost all aspects of their business.

The large-scale adoption of e-technology in Safety Significant Industries is inevitable. The nature of the market, its assets, processes and regulatory obligations however are likely to mean that the introduction will remain cautious and specialised.

Where a company sets out to employ an e-technology system to support a safety significant process and applies appropriate rigour, makes appropriate architectural decisions to support the use of COTS, records all necessary evidence and places the system well within a safety management framework then its use can be justified and controlled. There is no reason why, with the right planning and management, an e-technology system should not help a company to significantly improve business efficiency.

I believe the potential for 'context creep' when employing e-technologies is, however, significant. Only if the potential for creep is recognised by organisations and appropriately managed will the need for significant rework or, more dangerously, inappropriate use of such systems be avoided.

References

[Bishop, Bloomfield, Froome 2001] Bishop P, Bloomfield R E, Froome P K D: Justifying the use of software of uncertain pedigree (SOUP) in safety-related applications Report No: CRR336 HSE Books 2001 ISBN 0 7176 2010 7

[DTI 2000] Department of Trade and Industry: A Study of the Impact of e-Business on the UK Aerospace Sector, URN 00/1309, October 2000

[Glass 1997] Glass R: Software Runaways: Lessons Learned from Massive Software Project Failures – Pearson Education ISBN 013673443X.

[Jones, Bloomfield, Froome, Bishop 2001] Jones C, Bloomfield R E, Froome P K D, Bishop P: Methods for Assessing the Integrity of Safety-Related Software of Uncertain Pedigree (SOUP) Report No CRR337 HSE Books 2001 ISBN 0 7176 2011 5.

[Voas 1998] Voas J: An Approach to Certifying Off-the-Shelf Software Components IEEE Computer, June, 1998.

[Voas, Charron Miller 1996] Voas J, Charron F, Miller K: Tolerant Software Interfaces: Can COTS –based Systems be Trusted Without Them ? Proceedings of the 15th Int'l. Conference on Computer Safety, Reliability, and Security (SAFECOMP'96), Vienna, October 1996.

[Young 1999] Young E: Death March: The Complete Software Developers Guide to Surviving 'Mission Impossible' Projects Yourdon Computing Series - Prentice Hall ISBN 0130146595.

AUTHOR INDEX